# From Sunya to Algorithms

*Did India Invent Artificial Intelligence?*

# From Sunya to Algorithms

*Did India Invent Artificial Intelligence?*

Aryabhata, Brahmagupta, Panini, And Bhāskarāchārya to
The Machines That Think Today

*By*

Stephen Antony Venansious

notionpress.com

INDIA • SINGAPORE • MALAYSIA

# Preface

**Intelligence Before Machines**

Artificial Intelligence is often described as one of the greatest technological achievements of the modern age. News headlines, research laboratories, and corporate innovation centers frequently portray AI as something entirely new—a product of silicon chips, vast data centers, and complex machine learning algorithms created in the last few decades.

But this narrative raises an important question.

Where did the idea of algorithmic intelligence truly begin?

Did intelligence suddenly emerge when computers were invented, or does it belong to a much deeper intellectual lineage stretching across civilizations and centuries?

This book was born from that question.

While working in the fields of artificial intelligence, algorithmic governance, and digital safety systems, I repeatedly encountered ideas that seemed strangely familiar when viewed through the lens of ancient knowledge traditions. Concepts that modern AI researchers discuss today—formal rule systems, symbolic representation, recursive grammar, predictive modeling, and structured reasoning—had echoes in the intellectual heritage of ancient India.

The deeper I explored, the more striking the parallels became.

The concept of Śūnya (zero) fundamentally reshaped mathematics and made modern computation possible.

The grammatical system of Panini resembled a formal rule engine capable of generating language through ordered transformations.

The astronomical models of Aryabhata and the mathematical laws of Brahmagupta demonstrated algorithmic thinking centuries before the word *algorithm* itself entered the global vocabulary.

The mathematical and philosophical systems of Bhāskarāchārya revealed a tradition of predictive modeling and structured reasoning that strongly resembles modern computational frameworks.

These were not isolated discoveries.

They were parts of a broader intellectual ecosystem in which knowledge was treated as a system, not merely as a collection of facts. Mathematics, language, philosophy, astronomy, and ethics were deeply interconnected. Scholars were not only discovering truths about the world—they were designing methods for thinking about the world.

In many ways, that is precisely what artificial intelligence attempts to do today.

This book therefore does not argue that ancient India built computers or robots. That would be historically inaccurate and intellectually unnecessary. Instead, the central argument is far more interesting and far more important.

Artificial intelligence did not begin with machines.

It began with civilizations that learned how to encode intelligence into systems.

Ancient India was one of the earliest and most sophisticated examples of such a civilization.

Through centuries of scholarship, Indian thinkers developed formal approaches to language, mathematics, logic, and cosmology that transformed raw observation into structured knowledge. These systems were designed to be repeatable, teachable, and scalable—the very qualities that define algorithms today.

At the same time, Indian knowledge traditions also placed strong emphasis on ethics, responsibility, and the moral dimension of knowledge. Concepts such as Dharma, Ahimsa, and Karma framed intelligence not merely as a tool for power, but as a responsibility toward society and the natural world.

In the modern era, as artificial intelligence grows more powerful and pervasive, the global community is increasingly confronting the ethical implications of this technology. Governments are drafting regulations. Researchers are debating alignment and safety. International institutions are proposing frameworks for responsible AI.

Ironically, many of these ethical principles have existed for centuries within civilizational traditions that treated knowledge and morality as inseparable.

My own work in AI governance, algorithmic ethics, and digital child safety systems—including the development of the Algorethics AI Library, the Carlo Policy Enforcement platform, and the RoseShield digital child-safety ecosystem—has been deeply influenced by this realization.

Technology without ethics can amplify harm as easily as it can amplify progress.

As artificial intelligence reshapes the future of humanity, we must ask not only what machines can do, but what they should do.

The goal of this book is therefore twofold.

First, it seeks to reexamine the historical roots of algorithmic thinking, exploring how ancient Indian intellectual traditions contributed to the foundations of mathematics, computation, and structured reasoning.

Second, it seeks to reconnect technological intelligence with ethical intelligence, drawing inspiration from civilizational wisdom that understood knowledge as a force that must be guided by responsibility.

This work does not claim that any single civilization "invented" artificial intelligence in the modern technological sense. Intelligence evolves across cultures and centuries. Ideas travel, transform, and reappear in new forms.

However, when we look closely at the intellectual architecture underlying modern AI, we discover that many of its deepest conceptual foundations were anticipated by scholars who lived more than a thousand years ago.

Understanding this lineage is not merely an academic exercise.

It reminds us that the future of intelligence must be built not only on technological innovation, but on historical awareness, ethical responsibility, and respect for the wisdom of civilizations that came before us.

The story of artificial intelligence did not begin in Silicon Valley.

In many ways, it began with a simple yet profound idea from ancient India:

Śūnya — the power of structured emptiness — from which all computation emerges.

And from that idea grew a long and remarkable journey

from Śūnya to Algorithms.

—

Stephen Antony Venansious

Founder, Algorethics AI Library

Developer of Carlo ethics - The Ethical AI Firewall

www.carloethics.ai

Creator of RoseShield™ Digital Child-Safe Ecosystem

www.Childsafe.dev

✉ steve@algorethics.ai

# Dedication

This book is dedicated first to Almighty God, the source of all wisdom and truth, from whom every good gift, every discovery, and every insight ultimately flow

To my beloved parents,

Johny Nicolas Venansious and Kameela Venansious,

whose sacrifices, faith, and unwavering encouragement shaped my journey of learning and curiosity. Their love and discipline planted the seeds of inquiry that eventually led me to explore the deeper roots of knowledge across civilizations.

To my dear sisters,

Susan Mary Carmel and Regina Shanthi,

for their constant support, prayers, and belief in the pursuit of knowledge and truth.

To my beloved son,

Ebenizer Venansious Steaphen,

whose future represents the very reason this work exists. May the world he grows into build intelligent machines that remain guided by wisdom, ethics, and compassion.

And finally, to the ancient sages and scholars of Bharat —

to minds such as Aryabhata, Brahmagupta, Panini, and Bhāskarāchārya, and countless unnamed teachers of the Gurukul traditions — whose intellectual courage shaped humanity's understanding of numbers, language, logic, and systems thinking.

Their insights remind us that intelligence did not begin with machines, but with civilizations that sought harmony between knowledge, ethics, and the cosmos.

May their legacy guide the creation of ethical and safe artificial intelligence for future generations.

# About the Author

Stephen Antony Venansious is a technology entrepreneur, AI ethics researcher, and systems architect dedicated to building technologies that protect human dignity in the age of artificial intelligence.

He is the founder of the Algorethics AI Library, an open-source framework inspired by the *Rome Call for AI Ethics*, designed to embed transparency, accountability, and ethical safeguards into modern AI systems. His work focuses on the intersection of artificial intelligence, governance, algorithmic ethics, and digital safety, addressing emerging risks such as algorithmic bias, unsafe AI outputs, and the societal consequences of autonomous technologies.

Stephen is also the creator of RoseShield™, a privacy-first digital child-safety ecosystem designed to protect children across devices, homes, schools, and telecom networks. RoseShield detects and prevents online grooming, harmful conversations, manipulation, and AI-generated threats while preserving family privacy through on-device intelligence.

Complementing this initiative is ChildSafe.dev, a developer platform dedicated to building technologies that safeguard children from digital harm. The platform focuses on protecting young users from risks such as AI-driven manipulation, deepfakes, algorithmic bias, exploitation, and unsafe online interactions, while enabling ethical innovation in education and digital services.

Stephen has led research and development efforts in AI governance frameworks, algorithmic bias detection, ethical AI certification, and real-time policy enforcement systems. His Carlo Policy Enforcement as a Service (PEaaS) www.carloethics.ai platform enables continuous monitoring and enforcement of ethical standards within AI models and digital infrastructures.

In *From Śūnya to Algorithms: Did India Invent Artificial Intelligence?*, Stephen explores the historical lineage of algorithmic intelligence, tracing connections between the scientific achievements of ancient Indian scholars such as Aryabhata, Brahmagupta, Panini, and Bhāskarāchārya and the conceptual foundations of modern artificial intelligence.

Through this work, he argues that the roots of computation, formal systems, and algorithmic thinking can be traced to ancient Indian intellectual traditions that combined mathematics, philosophy, linguistics, and ethics into integrated systems of knowledge.

Stephen believes that the future of artificial intelligence must not be guided by technological power alone, but by civilizational wisdom, ethical responsibility, and respect for human dignity. His research and projects aim to build safe, ethical, and trustworthy AI systems that serve humanity rather than replace it.

🖂 **Contact:**

steve@algorethics.ai

# Contents

# Part I — The Shūnya Civilization (Foundations of Intelligence)

*Chapter 1*

# Before the Algorithm: The Civilization That Thought in Systems

The modern world tends to locate the birth of intelligence at the moment machines began to calculate. In popular discourse, intelligence is framed as a technological achievement an outcome of electricity, silicon, code, and computation. Artificial Intelligence, in this view, is the crowning achievement of modern science, born in laboratories, universities, and corporate research centers during the last century. Yet this framing reveals more about modern historical amnesia than about intelligence itself. Intelligence did not originate with machines. Machines are merely the most recent instruments through which intelligence expresses itself.

Long before the first computer executed a line of code, human civilizations were already solving complex problems of prediction, optimization, classification, memory, and reasoning. These problems were addressed not with hardware and software, but with intellectual systems frameworks of thought that allowed humans to model reality, anticipate outcomes, and adapt behavior. Among these civilizations, ancient Bharat stands out not simply for what it discovered, but for how it thought. This chapter explores a foundational premise of this book: that India was a civilization deeply oriented toward systemic intelligence, long before the concept of an algorithm was formally named.

To understand artificial intelligence, one must first understand intelligence without machines.

## Intelligence Before Machines

Intelligence, at its core, is the capacity to understand patterns, form abstractions, draw inferences, and act purposefully under conditions of uncertainty. Technology enhances intelligence, but it does not create it. A calculator accelerates arithmetic; it does not invent mathematics. Similarly, modern AI systems amplify pattern recognition and prediction; they do not originate the logic underlying those processes.

Ancient societies expressed intelligence through astronomy, agriculture, medicine, governance, language, architecture, and philosophy. These were not isolated domains but interconnected systems. Calendrical calculations influenced farming cycles. Astronomical observations shaped ritual timing and navigation. Linguistic precision enabled legal systems and philosophical debate. Intelligence was not siloed; it was integrated.

In Bharat, intelligence was understood as a relational phenomenon emerging from the interaction between mind, nature, and society. Knowledge was not merely accumulated; it was structured. Systems were designed to be repeatable, verifiable, and transmissible across generations. This emphasis on structure is critical. An unstructured insight may be brilliant, but it is not scalable. A structured insight becomes a system and systems are the foundation of computation.

What distinguishes Indian intellectual history is its early recognition that intelligence could be formalized. Ideas could be encoded into rules. Processes could be standardized. Reasoning itself could be examined, refined, and taught. This is the essence of algorithmic thinking, even in the absence of machines.

## Bharat as a Knowledge Civilization

Bharat was not defined solely by territory, dynasty, or empire. It was defined by its sustained commitment to knowledge creation, preservation, and transmission. While political powers rose and fell, intellectual traditions endured. This continuity was not accidental; it was the result of a civilizational architecture designed to protect and evolve knowledge.

Unlike civilizations where knowledge was concentrated within royal courts or restricted priesthoods, Indian civilization distributed knowledge across networks of teachers, students, monasteries, and learning centers. Knowledge moved through lineages rather than institutions alone. This decentralized structure made the civilization resilient to invasion, disruption, and political collapse.

Several defining features mark Bharat as a knowledge civilization:

- Knowledge was methodological, not merely descriptive. It focused on how to know, not just what to know.
- Knowledge was iterative, refined through commentary, debate, and revision.
- Knowledge was interdisciplinary, allowing insights from one field to inform another.
- Knowledge was ethical, tied to responsibility, purpose, and social harmony.

This orientation fostered abstract thinking at scale. Indian scholars did not merely observe phenomena; they sought underlying principles. Whether analyzing planetary motion, linguistic structure, or numerical systems, they consistently asked: What are the rules governing this domain?

Such a question is the starting point of every algorithm.

## Oral Intelligence and the Myth of Primitiveness

One of the most persistent misconceptions about ancient India is the assumption that oral traditions indicate intellectual primitiveness. In reality, India's oral knowledge systems represent one of the most sophisticated information-preservation frameworks ever developed by human civilization.

The Vedic corpus was transmitted orally for centuries possibly millennia with extraordinary fidelity. This was not the result of casual memorization but of highly formalized techniques designed to minimize error and detect corruption. These techniques functioned similarly to modern error-detection and correction mechanisms used in digital communication.

Oral transmission relied on redundancy, recursion, and pattern reinforcement. Texts were recited in multiple sequences and combinations to ensure that deviations could be identified. This made the oral system not weaker than writing, but in many ways stronger because it actively guarded against entropy.

When writing emerged as a supplementary medium, Indian intellectual culture did not abandon orality. Instead, it integrated the two. Written texts were deliberately concise, expressed in sutras compressed statements of rules or principles. These sutras required unpacking through oral explanation and commentary, ensuring that understanding accompanied memorization.

This balance between compression and interpretation mirrors modern computational design. High-level code abstracts complexity, but requires knowledge of underlying logic to be effective. Ancient Indian texts operated on the same principle. They assumed intelligence on the part of the reader or listener, not passive consumption.

## Education as a Living System: Gurukul, Takshashila, Nalanda

Education in ancient India was not standardized in the industrial sense. It was adaptive, relational, and system-oriented. Institutions such as the Gurukul, Takshashila, and Nalanda did not function as factories of information, but as ecosystems of learning.

## The Gurukul Model

The Gurukul was an immersive learning environment in which students lived with their teacher. This proximity allowed education to extend beyond formal lessons into daily life. Knowledge was transmitted not only through instruction, but through observation, dialogue, and practice.

Learning followed a feedback loop: instruction, application, correction, and refinement. Students progressed not by age or time, but by mastery. Ethical development was inseparable from intellectual growth. Intelligence was understood as incomplete without character.

From a systems perspective, the Gurukul operated as a closed-loop learning model, emphasizing continuous improvement an idea central to modern machine learning.

## Takshashila: Interdisciplinary Intelligence

Takshashila was among the world's earliest international centers of learning. Students from across Asia studied medicine, law, economics, warfare, linguistics, and philosophy. The absence of rigid disciplinary boundaries encouraged cross-pollination of ideas.

This interdisciplinary exposure enabled what modern researchers call transfer learning the ability to apply insights from one domain to another. Such flexibility is a hallmark of advanced intelligence, human or artificial.

## Nalanda: Distributed Knowledge at Scale

Nalanda represented an unprecedented scale of organized learning. Thousands of students and teachers lived, studied, and debated within a structured academic environment. Vast libraries preserved texts from multiple traditions. Debate was institutionalized as a method of evaluation.

Nalanda's system resembled a distributed knowledge network. Ideas were proposed, challenged, refined, and propagated through interaction. Authority rested not solely on hierarchy, but on reasoning and evidence. This environment cultivated collective intelligence a concept now explored in swarm intelligence and distributed AI systems.

## Structured Reasoning in Indian Thought

Perhaps the strongest evidence that ancient India thought in systems lies in its approach to reasoning itself. Indian philosophy did not treat logic as a mere intellectual exercise; it treated it as a practical tool for navigating reality.

The Nyaya school developed a formal theory of inference, identifying sources of knowledge, types of error, and rules of valid reasoning. Knowledge claims were evaluated not only on conclusions, but on the processes by which those conclusions were reached.

Reasoning was structured, not arbitrary. Arguments followed defined sequences. Assumptions were examined. Counterarguments were anticipated. Debate was not adversarial for its own sake; it was a method of refining truth.

Indian reasoning systems recognized uncertainty and context. Unlike rigid deterministic models, they allowed for probabilistic and conditional understanding. This flexibility is strikingly aligned with modern AI, which increasingly relies on probabilistic inference rather than absolute certainty.

Equally significant was India's comfort with abstraction. Concepts such as zero, infinity, and emptiness were not feared or avoided; they were explored. This willingness to engage with non-intuitive ideas expanded the boundaries of what could be modeled and computed.

Structured reasoning in India often included meta-rules rules about how rules operate. This recursive thinking appears in Panini's grammar, in mathematical formulations, and in philosophical inquiry. Recursion is not merely a technical feature; it is a cognitive leap. It allows systems to generate infinite expressions from finite rules.

Modern artificial intelligence relies heavily on this principle.

## Systems Thinking as a Civilizational Trait

What emerges from this examination is not a collection of isolated achievements, but a coherent pattern. Indian civilization consistently approached knowledge as a system. Language was systematized. Numbers were systematized. Reasoning was systematized. Education was systematized. Ethics were systematized.

This does not mean the civilization was rigid or mechanistic. On the contrary, its systems were designed to accommodate variation, interpretation, and growth. They were flexible frameworks rather than fixed formulas.

Systems thinking enabled scalability. Ideas could travel across regions and centuries because they were encoded into structures rather than dependent on individual genius alone. This is precisely what allows algorithms to scale today.

*Chapter 2*

# Śūnya: The Most Powerful Invention in Human History

Among all human inventions fire, language, agriculture, writing, the wheel, electricity none has altered the structure of reality itself as deeply as Śūnya. Zero did not merely improve calculation; it transformed how humanity thinks. It redefined absence, reorganized logic, enabled abstraction, and quietly became the backbone of modern civilization. Without Śūnya, there would be no algebra, no calculus, no physics as we know it, no computers, no artificial intelligence, and no digital universe.

Yet Śūnya is rarely acknowledged as an invention of this magnitude. Perhaps this is because it does not feel like an invention at all. It feels obvious, natural, inevitable. But this sense of inevitability is an illusion created by hindsight. For most of human history, zero did not exist not because people lacked intelligence, but because the idea required a radical philosophical leap. Śūnya demanded that humans accept a paradox: that nothing can be something, that absence can be formalized, and that emptiness can carry meaning.

This chapter explores Śūnya not only as a mathematical symbol but as a civilizational breakthrough one that emerged from deep philosophical soil, reshaped global knowledge systems, and ultimately made modern computing possible.

## What Is Śūnya? (Philosophical + Mathematical)

Śūnya is often translated simply as "zero," but this translation strips the concept of its depth. In Sanskrit, Śūnya means emptiness, void, openness, hollowness, or absence, but not in a negative sense. It does not imply annihilation or nonexistence. Instead, it refers to a state of potential, a condition in which form has not yet arisen or has temporarily dissolved.

Philosophically, Śūnya reflects a worldview in which reality is not composed solely of solid, independent entities. Instead, existence is relational, dynamic, and contingent. Things arise, change, and disappear. What we call "being" is inseparable from "non-being." Śūnya is the space that allows form to appear.

This understanding is deeply embedded in Indian philosophical traditions. In Vedantic thought, the unmanifest precedes the manifest. In Buddhism, Śūnyatā refers to the empty nature of all phenomena not meaning they do not exist, but that they lack fixed, independent essence. In Jain philosophy, reality is seen as multifaceted, with absence playing a role as significant as presence.

Mathematically, Śūnya became the symbol for zero, but its importance went far beyond representing "nothing." It allowed numbers to be positionally meaningful. Once zero existed, numbers could expand infinitely in both directions. Large quantities could be represented compactly. Abstract operations could be performed without physical reference.

Before zero, numbers were concrete. They counted objects. After zero, numbers became conceptual. They described states, relationships, possibilities, and limits.

This transition marks one of the most profound shifts in human cognition. With Śūnya, mathematics ceased to be merely a tool for trade or measurement and became a language capable of describing the universe itself.

Śūnya unified philosophy and mathematics into a single conceptual framework. It asserted that absence is not meaningless, that emptiness can be structured, and that reality includes what is not immediately visible.

## Zero vs Nothingness A Global Misinterpretation

One of the greatest intellectual misunderstandings in history is the equation of zero with nothingness. This confusion delayed the acceptance of zero across much of the world and continues to distort how people understand it today.

Nothingness implies absolute non-existence. Zero does not. Zero represents existence in a neutral or null state. It is not the absence of a value; it is a value that represents absence.

This distinction may appear subtle, but it has enormous consequences.

Many ancient civilizations had no problem recognizing absence in practical contexts. Merchants knew when a storehouse was empty. Astronomers knew when the sky showed no stars in a region. But they did not formalize absence as a numerical entity. Absence was situational, not symbolic.

Roman numerals provide a clear example. The Roman system had no zero. As a result, it could not easily express large numbers, negative values, or abstract calculations. Roman engineering was impressive, but Roman mathematics remained constrained because it lacked a way to encode absence structurally.

In medieval Europe, zero faced philosophical and theological resistance. Some scholars argued that God created the universe from nothing, and therefore nothingness should not be manipulated by humans. Others associated zero with chaos, void, or even moral danger. The idea that "nothing" could be written, calculated, and transformed seemed deeply unsettling.

This resistance was not technical; it was metaphysical. Accepting zero required accepting that absence can be part of reality, not merely the lack of it.

In contrast, Indian intellectual traditions had already spent centuries contemplating emptiness. The void was not feared. It was studied, respected, and integrated into metaphysical systems. As a result, when zero appeared in Indian mathematics, it did not feel alien. It felt inevitable.

The global misinterpretation of zero as nothingness shows how deeply philosophy shapes technology. Civilizations that could not philosophically accommodate emptiness could not mathematically operationalize it.

Even today, people struggle with zero conceptually. They find it difficult to grasp zero-based indexing in programming, null values in databases, or the idea that zero can be both something and nothing at once. These struggles echo ancient resistance.

Śūnya teaches that absence is not empty of meaning. It has structure, behavior, and consequences.

## Śūnya in the Upanishads & Mathematics

Long before zero appeared in written mathematical form, the philosophical foundations for Śūnya were already present in the Upanishads. These ancient texts explore the nature of reality, self, and existence with a depth that continues to astonish modern thinkers.

The Upanishads describe an ultimate reality that is formless, infinite, and beyond sensory perception. This reality is not nothingness in the nihilistic sense. It is fullness without form a paradox that closely mirrors the mathematical behavior of zero and infinity.

One famous Upanishadic declaration states:

"That is full. This is full. From fullness comes fullness. When fullness is taken from fullness, fullness remains."

This statement defies ordinary arithmetic logic, yet it aligns closely with advanced mathematical ideas involving limits, infinity, and zero. Removing infinity from infinity does not necessarily result in nothing. Similarly, zero can exist as a stable state without collapsing into nonexistence.

When Indian mathematicians such as Brahmagupta formalized zero in the 7th century, they did so within a philosophical environment already comfortable with paradox and abstraction. Brahmagupta treated zero as a number with defined operations. He explored its behavior in addition, subtraction, and multiplication with remarkable clarity.

This mathematical confidence did not arise in isolation. It was nurtured by centuries of metaphysical reflection on emptiness, form, and transformation. Mathematics in India did not invent zero from scratch; it translated philosophical insight into symbolic language.

The Upanishadic worldview allowed scholars to accept several ideas that were revolutionary for mathematics:

- That reality can be abstracted beyond physical objects
- That absence can be represented without contradiction
- That relationships matter more than substances

These principles are foundational to modern mathematics and science. They enabled algebraic thinking, symbolic manipulation, and theoretical modeling.

Śūnya thus stands at the intersection of metaphysics and mathematics. It is a reminder that great scientific advances often arise from deep philosophical clarity.

## Binary Logic Hidden in Śūnya

At the core of modern computing lies a deceptively simple system: binary logic. Everything digital text, images, sound, video, artificial intelligence is ultimately reduced to sequences of 1s and 0s. Presence and absence. On and off.

This binary system is not merely a technical convenience. It is the operational expression of Śūnya.

In computing, zero does not mean nothing is happening. It means a system is in a defined state. A circuit is closed. A signal is low. A condition evaluates to false. Zero is a decision state, not a void.

Binary logic works because it eliminates ambiguity. By reducing all possibilities to two states, systems become reliable, scalable, and predictable. But this reduction is only possible because zero exists as a meaningful, stable value.

Without zero, binary logic collapses. There is no false without zero. No off without zero. No baseline from which deviation can be measured.

Boolean algebra, which underpins all digital systems, relies entirely on zero and one. Every logical gate AND, OR, NOT depends on the interaction between presence and absence.

What makes this remarkable is that Śūnya anticipated this logic century before electronics existed. The philosophical insight that emptiness can be meaningful allowed future civilizations to encode absence as information.

Even advanced technologies such as neural networks and artificial intelligence rely on zero:

- Zero represents neutral weights
- Zero initializes learning systems
- Zero defines inactive neurons
- Zero serves as baseline comparison

The most complex machines ever built still depend on the same ancient idea: structured emptiness.

## Why Modern Computing Is Impossible Without Zero

To fully grasp why zero is indispensable to computing, one must look beyond hardware and into abstraction itself. Computers do not understand meaning. They process states. Every computation is a comparison, a transformation, or a transition between states.

Zero provides the reference point that makes these transitions intelligible.

Memory, for example, depends on zero. Empty memory is not undefined; it is initialized to zero. This allows systems to distinguish between stored data and random noise. Without a zero state, memory would be unstable and unreliable.

Programming logic depends on zero at every level. Conditional statements evaluate expressions as true or false one or zero. Loops terminate when counters reach zero. Errors are detected when values deviate from zero.

Even time in computing begins at zero. System clocks reset to zero. Epoch time starts from zero. Debugging often begins by checking whether a value is zero.

Data compression relies heavily on zero. Silence in audio, repetition in images, and redundancy in text are often represented by sequences of zeros. In this way, absence becomes efficiency.

Without zero:

- Positional number systems fail
- Binary logic cannot function
- Algorithms lose termination conditions
- Memory becomes chaotic
- Abstraction collapses

Modern computing is not built on numbers; it is built on states, and zero is the most fundamental state of all

## The Deeper Meaning of Śūnya

Śūnya is not merely a mathematical or technological concept. It carries a profound existential lesson. It teaches that absence is not failure, that emptiness is not deficiency, and that silence is not insignificance.

Creativity emerges from gaps. Insight arises in stillness. Innovation is born in uncertainty. The pause between thoughts is where understanding forms.

Civilizations that misunderstood zero lagged technologically. Individuals who fear emptiness often fear reflection, rest, and uncertainty. Yet growth requires space. Transformation requires dissolution. New forms arise only when old ones release.

Śūnya is not a symbol of lack. It is a symbol of possibility.

By recognizing zero, humanity recognized something even deeper: that reality is shaped as much by what is not there as by what is. This single insight gave rise to modern science, digital civilization, and the computational universe that now surrounds us.

Śūnya is not simply the most powerful invention in human history.

It is the quiet idea that made all other inventions possible.

*Chapter 3*

# Aryabhata: The First Algorithmic Mind

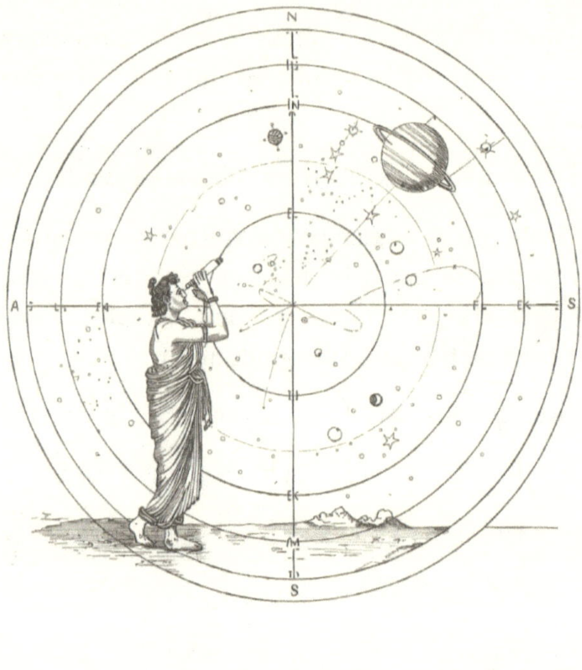


When modern scholars speak of algorithms, they often begin their story in medieval Europe or the early modern West, invoking names such as Al-Khwarizmi, Leibniz, or later, Turing. Yet this narrative overlooks a deeper and far older lineage of algorithmic thought one that emerged not from machines, but from the disciplined human mind. Long before computation became electronic, intelligence itself was formalized, structured, and operationalized. Among the earliest and most profound examples of this transformation stands Aryabhata, a mathematician and astronomer whose work reveals a mind thinking not just mathematically, but algorithmically.

Aryabhata did not invent machines that think, but he invented something more fundamental: methods that think. His contributions demonstrate that artificial intelligence, at its core, is not a technological event but a cognitive tradition one that began when humans learned to encode reasoning into repeatable systems. In this sense, Aryabhata can rightly be understood as the first known architect of algorithmic intelligence.

## Aryabhata's Life and Epoch: Intelligence Before Machines

Aryabhata was born in 476 CE, during a period of extraordinary intellectual vitality in the Indian subcontinent. This era, commonly associated with the Gupta civilization, witnessed significant advances in mathematics, astronomy, medicine, metallurgy, linguistics, and philosophy. Unlike many ancient societies where knowledge was restricted to priestly elites or royal courts, India had developed a distributed ecosystem of learning, supported by institutions such as Takshashila, Nalanda, and Kusumapura. These were not merely centers of instruction; they functioned as early research universities where debate, experimentation, and refinement of ideas were encouraged.

Aryabhata is most closely associated with Kusumapura, believed to be near present-day Patna. While historical records of his personal life are sparse, the intellectual clarity of his work speaks for itself. At the remarkably young age of approximately twenty-three, Aryabhata composed his seminal treatise, the Āryabhaṭīya. This text, written in succinct Sanskrit verses, contains within it an entire framework for astronomical calculation, mathematical reasoning, and predictive modeling.

The brevity of the Āryabhaṭīya is not a limitation; it is a feature. Aryabhata employed a sutra-based compression technique, where complex procedures were encoded into compact, memorizable forms. This was necessary in a culture where knowledge transmission was often oral, but it also demanded an exceptionally high level of abstraction. Each verse functioned not as explanation but as instruction a hallmark of algorithmic design.

Aryabhata lived in a civilization that had already separated process from personality. Knowledge was valued not because of who spoke it, but because it worked. This cultural foundation allowed Aryabhata to focus entirely on system-building rather than storytelling. His work does not persuade through rhetoric; it convinces through results.

## Algorithmic Astronomy: Turning the Cosmos into Computation

Astronomy in Aryabhata's time was not an abstract or speculative pursuit. It was a practical science essential for calendar-making, agriculture, navigation, and ritual timing. Accuracy was not optional it was foundational. Aryabhata approached this challenge not by appealing to mythological explanations, but by constructing mathematical models of celestial motion.

He treated the Sun, Moon, and planets as entities governed by rules that could be described numerically. His methods allowed scholars to calculate planetary positions at any given time, predict eclipses, and determine the length of the year with impressive precision. What distinguishes Aryabhata's astronomy from earlier observational traditions is his insistence on procedural calculation.

Aryabhata did not simply state astronomical facts. He provided step-by-step methods algorithms that could be executed repeatedly to obtain results. These methods relied on initial parameters, mathematical operations, and systematic iteration. In modern terms, Aryabhata was building computational pipelines.

This approach reflects a deep understanding of abstraction. The sky was no longer a collection of divine symbols; it became a data-generating system. Time became a variable. Motion became a function. Prediction became the goal.

Importantly, Aryabhata understood the limitations of precision. His calculations often involved approximations, revealing an early awareness of what modern science calls model error. Rather than seeking absolute perfection, he sought usable accuracy. This mindset is identical to that of modern AI systems, which prioritize predictive performance over theoretical exactness.

In this way, Aryabhata transformed astronomy into an early form of computational science, centuries before the existence of computing machines.

## Trigonometry as a Computational Language

One of Aryabhata's most enduring contributions lies in his formalization of trigonometry as a computational tool. While geometric reasoning existed in many ancient cultures, Aryabhata's innovation was to translate geometric relationships into numerical functions that could be calculated, tabulated, and reused.

He introduced the sine function (jya) and provided a table of sine values calculated at regular angular intervals. This table was not decorative; it was an operational resource designed for repeated use in astronomical computation. The very existence of such tables indicates a sophisticated understanding of reusability and standardization, core principles of algorithmic systems.

Trigonometry, as Aryabhata conceived it, acted as a bridge between continuous reality and discrete calculation. Circular motion, inherently smooth and infinite, was broken down into manageable numerical segments. This is conceptually identical to how modern digital systems sample continuous signals and convert them into discrete data.

In effect, Aryabhata created an early form of numerical encoding. The heavens were translated into numbers, and numbers into procedures. Once encoded, these procedures could be applied mechanically by trained scholars, much like software executed by hardware.

This transformation of geometry into computation is a critical step in the evolution of artificial intelligence. AI systems do not perceive reality directly; they operate on representations. Aryabhata was among the first to demonstrate how such representations could be mathematically constructed and systematically manipulated.

## Place Value and Positional Logic: The Architecture of Thought

Underlying Aryabhata's entire mathematical system is the place value notation that had already matured in Indian mathematics. While Aryabhata may not have explicitly theorized zero in philosophical terms, his work fully depends on the logic of positional notation and the use of śūnya as a placeholder.

This system represents one of the most powerful cognitive technologies ever invented. Place value notation allows a finite set of symbols to express infinite quantities. It enables scaling, compression, and abstraction all essential for complex computation.

Aryabhata's algorithms rely on positional logic for efficiency. Calculations involving large astronomical cycles would be practically impossible without such a system. By assuming positional notation as a given, Aryabhata was effectively programming within a highly advanced numerical architecture.

Modern computing operates on the same principle. Binary code, despite its apparent simplicity, is nothing more than positional notation applied at scale. Every digital system, including artificial intelligence, rests on this foundation.

Aryabhata's ease with positional logic suggests that Indian mathematics had already internalized a form of symbolic intelligence. Numbers were no longer tied to physical objects; they were abstract entities that could be manipulated according to rules. This abstraction is the defining feature of algorithmic thinking.

## The Rotating Earth and Predictive Models of Reality

Perhaps one of Aryabhata's most intellectually courageous contributions was his assertion that the Earth rotates on its axis. This idea directly challenged intuitive perception and prevailing cosmological beliefs. Yet Aryabhata arrived at this conclusion not through speculation, but through model optimization.

By assuming a rotating Earth, Aryabhata simplified the explanation of apparent stellar motion. The heavens did not need to spin endlessly; instead, the observer was moving. This shift represents a profound cognitive leap the willingness to separate observed phenomena from underlying structure.

This distinction lies at the heart of modern science and artificial intelligence. AI systems must infer hidden causes from visible data. Aryabhata demonstrated this same capability by identifying the true source of celestial motion.

His eclipse calculations further reinforce this predictive mindset. By modeling the relative positions of the Sun, Moon, and Earth, Aryabhata could forecast eclipses with remarkable accuracy. These predictions were forward-looking, not retrospective. They functioned as simulations of future states, a key feature of intelligent systems.

In modern terms, Aryabhata was running mental simulations of physical reality. He created internal models, adjusted parameters, and evaluated outcomes. This is precisely how modern AI systems perform forecasting and planning.

## Why Aryabhata Was an Algorithm Designer

To describe Aryabhata as an algorithm designer is not metaphorical it is structurally accurate. His work exhibits all the defining characteristics of algorithmic systems.

His methods were rule-based rather than intuitive. They were general rather than case-specific. They were designed for repetition, verification, and correction. Most importantly, they separated method from outcome, allowing knowledge to be scaled beyond the individual thinker.

Several features of Aryabhata's approach stand out:

- He encoded complex reasoning into compact, executable forms
- He prioritized procedures over explanations
- He designed systems that produced predictions, not just descriptions
- He treated natural phenomena as variables within formal models

These are the same principles that govern modern algorithm design.

Aryabhata's audience trained scholars functioned as the execution engine. The verses were the code. The results were the output. This human-based computation may seem distant from modern machines, but the underlying logic is identical.

What Aryabhata ultimately achieved was the externalization of intelligence. Thought was no longer confined to the mind; it was embedded in structured systems that others could learn, execute, and extend. This is the true beginning of artificial intelligence.

## Aryabhata's Enduring Relevance in the Age of AI

In the twenty-first century, as humanity grapples with machines that can learn, predict, and generate knowledge, Aryabhata's work feels unexpectedly contemporary. His emphasis on abstraction, modeling, prediction, and efficiency aligns perfectly with the goals of modern AI.

Artificial intelligence did not emerge suddenly with computers. It emerged when humans learned to formalize thought into systems. Aryabhata stands among the earliest known masters of this art.

He reminds us that intelligence is not a product of hardware, but of methodological discipline. Long before silicon, long before electricity, and long before the word "algorithm" existed, Aryabhata was already thinking in ways that define intelligent systems today.

To recognize Aryabhata as the first algorithmic mind is not to diminish modern achievements. It is to place them within a longer, richer history one that begins not in laboratories, but in the disciplined imagination of ancient India.

*Chapter 4*

# Brahmagupta: Zero, Negatives & Computational Laws

If Aryabhata represents the algorithmic awakening of Indian mathematics, then Brahmagupta represents its formalization. Where earlier thinkers demonstrated extraordinary computational intuition, Brahmagupta insisted on something more radical: that abstraction itself must obey explicit, written laws. His contribution was not merely numerical; it was epistemological. He transformed mathematics from practiced technique into a rule-governed system, and in doing so, laid intellectual groundwork that modern computation still depends upon.

Brahmagupta was born in 598 CE in Bhinmal, a major center of learning in western India. By his time, Indian mathematics had already surpassed most of the world in sophistication. The place-value system was mature, trigonometric methods were advanced, and astronomical modeling was deeply algorithmic. Yet even within this fertile environment, something crucial was missing. Zero existed, but it lacked a fully articulated behavioral framework. Negative quantities were understood intuitively, but not systematically. Error correction existed in practice, but not yet as a formal concern.

Brahmagupta confronted all of these gaps directly. His masterpiece, the Brahmasphuṭasiddhānta, written in 628 CE, is not simply a mathematical treatise; it is a declaration that numbers, even when abstract or paradoxical, must obey consistent laws. This insistence marks a turning point not only in mathematics, but in the long evolution of computational reasoning.

## Formal Rules of Zero

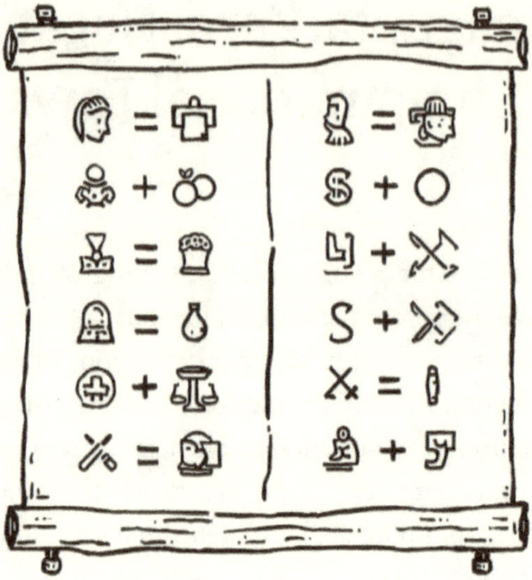

Before Brahmagupta, zero (śūnya) was present in Indian thought as both symbol and concept. Philosophically, it carried immense weight in metaphysics, cosmology, and logic. Mathematically, it functioned primarily as a placeholder within the decimal place-value system. What it lacked was something essential for any computational system: formal operational rules.

Brahmagupta recognized that without such rules, zero would remain fragile useful but unstable, powerful but undefined. A system that cannot predict how its foundational elements behave under transformation cannot scale, and it cannot be trusted.

In the Brahmasphuṭasiddhānta, Brahmagupta explicitly defines how zero behaves under arithmetic operations. He states, unambiguously, that when zero is added to a number, the number remains unchanged, and when a number is subtracted from itself, the result is zero. He further clarifies that multiplication by zero annihilates quantity, collapsing it into nothingness.

These statements may appear trivial to modern readers, but they represent a monumental shift. Brahmagupta was not merely describing patterns; he was legislating behavior. He treated zero as a lawful entity within a closed system of rules, rather than as an exception or philosophical curiosity.

This move parallels a foundational principle in computer science: all states must be defined. Undefined behavior is the enemy of computation. In modern programming languages, even null values and empty states are explicitly managed, precisely because leaving them ambiguous leads to system failure. Brahmagupta's zero is an early expression of this same logic.

What is most striking is his confidence. Zero was not treated cautiously or reverently. It was treated operationally. Brahmagupta did not ask whether zero should exist; he assumed it must, and then asked how it behaves.

This shift from ontological doubt to operational clarity is one of the deepest intellectual transitions in human history.

## Operations on Nothingness

Brahmagupta's boldness did not stop with basic arithmetic. He ventured into territory that even modern mathematics handles carefully: operations involving nothingness itself.

He defined rules for multiplication and division involving zero, fully aware that these operations touched the boundaries of logical coherence. While some of his conclusions particularly regarding division by zero do not align with modern mathematical conventions, the importance of his effort lies not in perfection, but in system completion.

Brahmagupta believed that a mathematical system should not avert its eyes from paradox. Instead, it should confront ambiguity directly and attempt resolution. This is a profoundly computational mindset. In algorithm design, incomplete specifications are more dangerous than incorrect ones, because they produce unpredictable behavior.

His attempt to define division by zero reflects this philosophy. Rather than declaring it forbidden or leaving it undefined, Brahmagupta attempted to describe its behavior within the system available to him. Even when later mathematicians refined or corrected these rules, the impulse toward completeness remained.

Equally transformative was his formal treatment of negative numbers. Using the intuitive concepts of fortune (positive) and debt (negative), Brahmagupta constructed a coherent arithmetic of opposites. He explained how debts interact with fortunes, how debts compound, and how opposites neutralize each other.

This was not merely pedagogical cleverness. It was a semantic model, grounding abstract operations in lived economic reality. Modern computational systems rely on similar mappings, translating machine operations into metaphors humans can reason about: memory, stacks, queues, loss, reward.

By systematizing negatives, Brahmagupta extended the numerical universe into a bidirectional space. This conceptual expansion is indispensable to modern science and artificial intelligence. Optimization algorithms depend on gradients that move through negative space. Neural networks minimize error by descending into negative values. Loss functions, penalties, and corrective signals all presuppose a mathematical world where negativity is not only allowed, but essential.

Brahmagupta did not merely tolerate negatives. He normalized them.

## Early Formal System Theory

What distinguishes Brahmagupta most sharply from many ancient mathematicians is his implicit understanding of mathematics as a formal system. A formal system is not defined by its subject matter, but by its structure: symbols governed by transformation rules that yield predictable outcomes.

Brahmagupta's work satisfies this definition in a remarkably modern way. He defined symbols (numbers, zero, negatives), operations (addition, subtraction, multiplication, division), and rules that applied universally across cases. His rules were abstract, not tied to specific geometric diagrams or physical measurements. They could be applied anywhere, by anyone who learned them.

This abstraction is the essence of computation.

An algorithm does not care whether it processes stars, coins, or words. It operates on symbols according to rules. Brahmagupta's arithmetic of zero and negatives functions exactly this way. It is context-independent, portable across domains.

This is why his ideas traveled so effectively across cultures. Techniques rooted in local practices often fail to migrate. Formal systems migrate easily. Once translated, they retain their power.

There is also a deeper philosophical implication. By incorporating zero and negatives into a lawful system, Brahmagupta implicitly argued that absence, opposition, and imbalance are computable phenomena. Reality, in this view, is not only what is present, but also what is missing, inverted, or opposed.

Modern artificial intelligence depends fundamentally on this assumption. AI systems do not learn from success alone; they learn from error. They adjust weights based on deviation, not alignment. In this sense, intelligence emerges not from affirmation, but from correction.

Brahmagupta's mathematics made such thinking intellectually legitimate.

## Error Correction in Ancient Computation

One of the most overlooked aspects of Brahmagupta's work is his sensitivity to error. Ancient astronomy was computationally demanding. Predicting planetary positions required long chains of calculation, where small inaccuracies could produce catastrophic deviations.

Brahmagupta did not assume perfection. He expected error, and he designed methods to manage it. He corrected earlier astronomical constants, refined predictive models, and revised inherited assumptions when they failed to match observation.

This attitude reflects an early understanding of what modern science calls iterative refinement. No model is final. Validity is provisional, dependent on accuracy and consistency.

In computational terms, this is the foundation of feedback loops. Modern machine learning systems are trained through repeated cycles of prediction, evaluation, and correction. Error is not a failure state; it is the engine of improvement.

Brahmagupta's willingness to revise rules and recalibrate values demonstrates a dynamic view of knowledge. Mathematics was not a static revelation but a living system, capable of self-correction.

This stands in contrast to later traditions that treated mathematical axioms as immutable truths. Brahmagupta's approach was closer to modern engineering than medieval scholasticism. He valued functionality over dogma.

Such an attitude is essential for artificial intelligence, where systems must adapt to changing data, environments, and objectives. Static intelligence fails. Adaptive intelligence survives.

## Influence on Arab Mathematical Systems

The global importance of Brahmagupta becomes unmistakable when we trace his influence beyond India. During the Abbasid Caliphate, particularly in the ninth century, Indian mathematical and astronomical texts were translated into Arabic as part of the great intellectual movement centered in Baghdad.

Brahmagupta's work entered the Islamic world through texts known collectively as Sindhind, derived from Sanskrit sources. These translations carried not only numerical techniques, but conceptual frameworks: zero, place value, and formal operational rules.

Arab mathematicians recognized the power of these ideas immediately. Scholars such as Al-Khwarizmi integrated Indian numerical systems into their own work, using them to advance algebra, astronomy, and computation. The very word "algorithm" derives from Al-Khwarizmi's name, yet the conceptual machinery he employed was deeply indebted to Indian mathematics.

As knowledge moved westward into Europe through Latin translations, attribution grew increasingly blurred. Indian numerals became "Arabic numerals." Zero lost its Sanskrit name. Rule-based arithmetic became absorbed into European mathematics with little acknowledgment of its origins.

This was not initially a deliberate act of erasure, but over time, it hardened into a narrative that credited Europe with discoveries built on

Indian and Islamic foundations. Brahmagupta's role faded from popular memory, even as his ideas continued to power global computation.

Recognizing this lineage does not diminish later contributions. It restores continuity. Artificial intelligence did not emerge suddenly in Silicon Valley. It stands on centuries of abstraction, formalization, and intellectual courage, much of which originated in India.

## Brahmagupta and the Foundations of Artificial Intelligence

Brahmagupta did not invent artificial intelligence. But he helped invent the conditions under which artificial intelligence is possible.

AI depends on:

- Zero as a neutral reference point
- Negative values as error signals
- Formal rules that govern transformation
- Systems resilient to correction and refinement

All of these were articulated, normalized, and defended by Brahmagupta more than a millennium ago.

Modern neural networks operate in numerical spaces dense with positives and negatives, converging toward zero-loss states. Optimization algorithms move by subtraction, not addition. Intelligence, in machines, is defined not by what they know, but by how they correct themselves.

Brahmagupta's greatest contribution, therefore, is not a formula. It is a worldview: that intelligence can be expressed through lawful abstraction, that even nothingness can be computed, and that systems grow stronger by confronting their own errors.

In the silent arithmetic beneath every intelligent machine, Brahmagupta's legacy still runs.

Not as memory.

But as law.

# Chapter 5

# Panini: The World's First Formal Programming Language

When the history of computation is told, it is usually framed as a modern Western achievement. The story often begins with mathematical logic in Europe, accelerates with mechanical calculators, and culminates in digital computers and artificial intelligence. Language, in this narrative, becomes something machines learned to handle only recently, through data, statistics, and silicon. Yet this version of history omits a profound and uncomfortable truth: the formalization of language as a computable, rule-governed system did not begin with computers at all. It began more than two millennia ago, in ancient India, with a grammarian named Pāṇini.

Pāṇini's Aṣṭādhyāyī is frequently described as a grammatical treatise, but this description dramatically understates its significance. It is not merely a record of linguistic conventions. It is a formal system, a complete and self-contained architecture that takes symbolic inputs and produces valid outputs through an ordered sequence of rules. In modern terms, it behaves like a programming language, complete with syntax, execution order, meta-rules, recursion, and error handling. What makes this extraordinary is not only its age, but its sophistication. Many of the ideas that underpin modern programming language theory appear in Pāṇini's work in fully developed form.

To call Pāṇini the creator of the world's first formal programming language is not metaphorical praise. It is a technically defensible claim, supported by structural, logical, and functional parallels that are now widely acknowledged in linguistics, computer science, and artificial intelligence research. This chapter explores that claim in depth, showing how the Aṣṭādhyāyī operates as a rule-based system, how it employs meta-rules, recursion, and compression, how it anticipates modern compiler design, how it embodies the core ideas of natural language processing long before the term existed, and why contemporary AI language models are Paninian in their very nature.

## The Aṣṭādhyāyī as a Rule-Based System

The Aṣṭādhyāyī consists of approximately 4,000 sutras, arranged across eight chapters. These sutras are not explanations or examples. They are instructions. Each sutra is deliberately concise, stripped of anything unnecessary, and designed to function within a larger system of interdependent rules. In isolation, a sutra may appear cryptic. In execution, it becomes precise.

Pāṇini's fundamental insight was that language could be generated from a finite set of abstract elements using a finite set of rules. He began not with words, but with roots and bases, which function as symbolic primitives. These primitives are then transformed through a controlled sequence of operations. Affixes are added, sounds are modified, substitutions occur, and constraints are enforced. At every step, the system checks conditions to determine which rule applies.

This is the hallmark of a rule-based computational system. The grammar does not list all valid Sanskrit words or sentences. That would be impossible. Instead, it defines a procedure by which valid forms can be generated. This generative approach is central to modern computing. A compiler does not store every possible program; it applies rules to interpret and transform code. Pāṇini's grammar operates in exactly the same conceptual manner.

Crucially, the rules are ordered. Pāṇini specifies which rules take precedence and under what circumstances. When multiple rules could apply, the system does not collapse into ambiguity. Instead, it resolves conflicts deterministically. This ensures that the same input always produces the same output. Determinism is a foundational property of formal systems, and Pāṇini enforced it with extraordinary discipline.

The Aṣṭādhyāyī also separates data from operations, another hallmark of computational thinking. Roots and stems function as data; sutras function as operations. This separation allows the same operations to be reused across countless derivations. Modern programming languages rely on the same principle to achieve scalability and abstraction.

Seen this way, Pāṇini did not merely analyze Sanskrit. He encoded it. He transformed a living language into an executable system. This act represents one of the earliest known examples of treating human cognition itself as something that could be formalized, structured, and processed.

## Meta-Rules, Recursion, and Compression

One of the strongest arguments for viewing the Aṣṭādhyāyī as a programming language lies in its use of meta-rules. These are rules that govern how other rules behave. Without them, the system would be unmanageable. With them, it becomes elegant and coherent.

Meta-rules in Pāṇini's system determine issues such as scope, inheritance, and precedence. They answer questions like: when two rules conflict, which one wins? When does the effect of a rule begin and end? When is a general rule overridden by a specific one? These questions are not peripheral; they are central to the functioning of any formal language.

In modern programming, similar concerns are handled through language specifications and execution models. Pāṇini anticipated these concerns and addressed them directly, embedding solutions into the grammar itself. This level of self-referential design is extraordinarily rare in ancient systems of thought.

Recursion is another defining feature. Pāṇini's rules often refer back to categories or structures defined earlier, allowing the same pattern to be applied repeatedly. This enables a compact set of rules to generate immense complexity. Recursion allows language to be both finite and infinite at the same time, finite in its rules, infinite in its expressions.

From a computational standpoint, recursion is not optional. It is essential for modeling hierarchical structures and nested dependencies. Modern programming languages rely on it extensively. Pāṇini's grammar demonstrates an intuitive grasp of this principle centuries before formal logic articulated it.

Compression is where Pāṇini's brilliance becomes almost staggering. The sutra format achieves an extraordinary density of information. Entire operations are encoded in a handful of syllables. This compression was not merely stylistic. It was functional. It allowed the grammar to be memorized, transmitted orally, and executed mentally without loss of precision.

In modern terms, Pāṇini optimized for:

- Minimal instruction length
- Maximum reuse of rules
- High information density
- Zero redundancy

This is exactly what modern programming language designers strive for when creating efficient, expressive languages. Pāṇini achieved it without machines, using only human cognition as the execution environment.

## Paninian Grammar and Modern Compiler Design

The structural similarities between Paninian grammar and modern compiler design are not coincidental. They arise because both are attempting to solve the same fundamental problem: how to transform symbolic input into structured output according to formal rules.

A modern compiler processes source code through several stages. First, it breaks the input into tokens. Then it checks whether those tokens conform to syntactic rules. Next, it applies semantic constraints. Finally, it transforms the input into a different representation.

Pāṇini's system follows a comparable pipeline. His classification of sounds and roots functions like lexical analysis. His rules for combining elements correspond to syntactic parsing. His constraints on meaning and usage resemble semantic checks. The final output is a valid linguistic form.

One of the most advanced aspects of Paninian grammar is its handling of context sensitivity. Many modern programming languages struggle with context-sensitive constructs because they require awareness of surrounding elements. Sanskrit, as formalized by Pāṇini, is governed by a highly context-sensitive system, and yet it remains computationally manageable due to the precision of the rules.

This is one reason why Paninian grammar has attracted the attention of computer scientists working in formal language theory. The Aṣṭādhyāyī demonstrates that context sensitivity can be handled systematically without chaos, provided the rule system is well-designed.

The implication is profound. It suggests that the intellectual foundations of compiler design did not emerge suddenly in the industrial age. They were anticipated in ancient India through a deep engagement with language as a formal system.

## NLP Before Natural Language Processing

Natural Language Processing is often presented as a triumph of modern computing. Machines, we are told, have finally learned to handle human language. Yet this framing obscures the fact that language had already been rigorously formalized long before computers existed.

Pāṇini treated language as something that could be modeled, generated, and validated. He identified minimal units, defined transformation rules, and constrained outputs. These are precisely the steps that NLP systems perform today, albeit with different tools.

The difference between Pāṇini and modern NLP is not conceptual but technological. Pāṇini relied on explicit rules and human execution. Modern NLP often relies on statistical learning and machine execution. Both approaches assume that language has an underlying structure that can be captured.

In fact, one of the challenges in modern NLP is dealing with ambiguity and irregularity. Sanskrit, as formalized by Pāṇini, minimizes these problems by enforcing strict derivational paths. Each form has a traceable origin. This makes it particularly attractive for computational modeling.

It is no coincidence that some early NLP researchers turned to Paninian grammar when seeking alternatives to Western linguistic models. They recognized that Pāṇini had already solved many of the problems they were struggling with.

## Why AI Language Models Are Paninian by Nature

At first glance, large language models may seem entirely different from Pāṇini's deterministic system. They are probabilistic, data-driven, and opaque. Pāṇini's grammar is explicit, rule-based, and transparent. Yet beneath these differences lies a shared foundation.

Both systems assume that language is generative. Neither stores all possible sentences. Both rely on internal structures that allow new expressions to be produced. Both operate on symbolic units and relationships between them.

AI language models break text into tokens and learn how these tokens relate. Pāṇini broke language into roots and affixes and defined how they relate. The tools differ, but the conceptual move is the same.

Moreover, AI models rely on architecture-defined constraints. They do not generate arbitrary sequences; they generate sequences that conform to learned structure. Pāṇini's system imposed constraints explicitly. AI systems learn them implicitly.

This convergence suggests that modern AI has rediscovered, through data and computation, principles that were articulated explicitly in ancient India. It is not that AI copied Pāṇini. It is that both arise from the same recognition: language is structured, and intelligence emerges from understanding that structure.

# *Chapter 6*

# Bhāskarāchārya & Early Predictive Modeling

The story of Bhāskarāchārya's daughter Lilavati offers a striking entry point into the philosophy of prediction. According to tradition, Bhāskara calculated with extraordinary precision the most auspicious moment for her marriage and devised a water clock to ensure that the ceremony would occur at exactly the right instant. Yet, in a moment of innocent curiosity, Lilavati leaned over the device, and a pearl from her ornament slipped into the vessel, obstructing the flow of water. The clock was delayed, the precise moment passed, and the carefully computed alignment of fate was lost. This episode, whether historical or symbolic, reveals a profound truth: even the most exact systems of calculation remain vulnerable to small, unforeseen disturbances. It is within this tension between precision and unpredictability that the deeper question of predictive intelligence emerges.

When modern artificial intelligence systems forecast weather patterns, optimize logistics, predict stock prices, or simulate planetary motion, they rely on mathematical frameworks that treat the future as something that can be modeled rather than guessed. Prediction, in the modern sense, is not prophecy it is structured inference. Long before computers processed data at electronic speed, Indian mathematicians had already begun asking the most important question behind predictive intelligence: Can the future be calculated from the present?

Among the towering figures who engaged deeply with this question was Bhāskarāchārya, also known as Bhāskara II, a 12th-century mathematician and astronomer whose work represents one of the highest achievements of pre-modern mathematical intelligence. Bhāskara did not merely solve equations; he modeled change. He did not merely observe celestial movement; he simulated it. His work reveals a sophisticated understanding of rates, limits, optimization, and forecasting concepts that sit at the very heart of modern predictive analytics and artificial intelligence.

This chapter explores Bhāskarāchārya not simply as a historical mathematician, but as an early architect of predictive reasoning. Through his seminal works Lilavati and Bijaganita, and through his astronomical treatises, Bhāskara demonstrated that reality could be translated into formal models capable of anticipating outcomes. In doing so, he contributed to a lineage of thought that ultimately made machine prediction possible.

## Lilavati & Bijaganita: Mathematics as a Living System

Bhāskarāchārya's two most famous mathematical works, Lilavati and Bijaganita, are often introduced as textbooks of arithmetic and algebra. Such descriptions, while not incorrect, significantly underestimate their depth. These texts were not mere compilations of techniques; they were carefully structured systems of reasoning designed to train the mind to think relationally, abstractly, and predictively.

Lilavati, focused primarily on arithmetic, is written in verse form. Its problems are presented through narratives involving daily life commerce, travel, distribution, and measurement. At first glance, these may appear elementary. In reality, they encode mathematical relationships that require the reader to translate real-world situations into symbolic structures. This translation process is foundational to all predictive modeling.

Bijaganita, which deals with algebra, advances this abstraction further. Here, Bhāskara works with unknown quantities, equations, and operations that manipulate relationships rather than concrete numbers. Algebra, by its nature, is predictive: it allows one to solve for unknowns based on known constraints.

What distinguishes Bhāskara's approach is his emphasis on general methods rather than isolated solutions. Problems are not solved once; they are solved in principle. The reader is taught how to derive solutions systematically, enabling the application of the same logic to new situations. This emphasis on method over instance is precisely what separates intelligent systems from rote calculation.

In modern terms, Lilavati and Bijaganita function as early training manuals for algorithmic thinking. They teach the reader how to structure problems, define variables, apply constraints, and arrive at outcomes through repeatable procedures.

## Mathematics as Modeling, Not Memorization

A critical feature of Bhāskarāchārya's work is that it treats mathematics as a model of reality, not merely a tool for calculation. Each problem implicitly asks the reader to build a mental model: identify inputs, understand relationships, and compute outputs. This is the same cognitive process used in modern data science and machine learning.

Bhāskara consistently emphasized reasoning over rote procedure. He explained why methods worked, not just how to apply them. This reflects a deep understanding that intelligence lies not in answers, but in the structures that produce answers.

Even in arithmetic problems, Bhāskara often dealt with rates of change, proportional reasoning, and dependency between variables. These are not static computations; they are dynamic relationships. A change in one quantity affects another. Such thinking naturally leads toward prediction.

Modern predictive systems operate on exactly this principle. Whether modeling population growth, financial markets, or neural activity, the core task is to represent how variables interact over time. Bhāskarāchārya's mathematics trained minds to do precisely this without machines.

## Early Calculus Concepts: Understanding Change

Perhaps the most remarkable aspect of Bhāskarāchārya's legacy is his treatment of ideas that closely resemble calculus, centuries before its formal development in Europe. While Bhāskara did not invent calculus as it is known today, he clearly understood many of its foundational concepts.

One of these concepts is the idea of instantaneous change. Bhāskara analyzed how quantities change at specific moments, not just over fixed intervals. This is evident in his astronomical calculations, where he examined the changing velocity of planets and sought to identify moments of maximum or minimum motion.

In one notable example, Bhāskara described how the rate of change of a sine function approaches zero at its maximum value. This insight reflects an intuitive grasp of derivatives and extrema concepts central to calculus and optimization.

Bhāskara also worked with notions of infinitesimal change, even if not formalized symbolically. He understood that smooth curves could be approximated by very small increments, allowing complex motion to be analyzed step by step. This incremental reasoning is fundamental to numerical methods used in modern simulations.

These ideas matter deeply for artificial intelligence. Learning algorithms rely on gradients rates of change to optimize models. Gradient descent, one of the most widely used optimization techniques in machine learning, is built entirely on calculus. Bhāskarāchārya's intuitive understanding of change places him firmly within the intellectual ancestry of such methods.

## Optimization & Limit Theory

Optimization is the process of finding the best possible outcome under given constraints. Whether minimizing error in a neural network or maximizing efficiency in logistics, optimization lies at the heart of modern AI. Bhāskarāchārya engaged deeply with this idea long before it was formalized mathematically.

In his astronomical work, Bhāskara sought to determine points of maximum and minimum planetary motion. This required identifying limits points beyond which a quantity could not increase or decrease. His reasoning shows an awareness that systems have boundaries and that meaningful prediction requires understanding those boundaries.

Bhāskara's treatment of limits was not abstract speculation. It was practical. To predict eclipses, planetary conjunctions, or seasonal changes, one must understand cyclical behavior and turning points. These are essentially optimization problems embedded in natural phenomena.

Bhāskara also demonstrated an understanding of approximation. He recognized that perfect precision is often unattainable, and that useful prediction depends on sufficiently accurate estimates. This pragmatic approach aligns closely with modern computational modeling, where approximation is a necessity rather than a flaw.

Optimization in Bhāskara's work was always contextual. Solutions were evaluated not only mathematically, but in terms of their alignment with observed reality. This emphasis on validation anticipates the modern practice of testing models against data.

## Forecasting & Astronomical Simulations

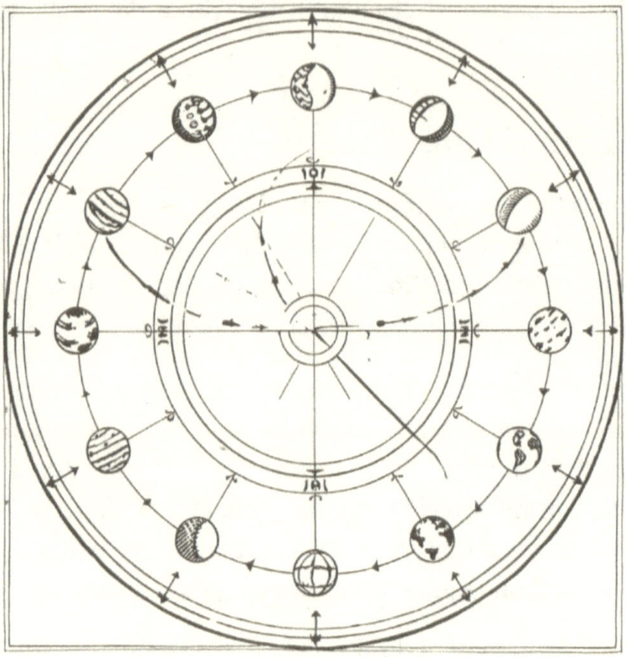

Astronomy was the ultimate testbed for predictive intelligence in the ancient world. The movement of celestial bodies is complex, cyclical, and sensitive to initial conditions. Predicting eclipses or planetary positions requires the ability to simulate dynamic systems over time.

Bhāskarāchārya excelled in this domain. His astronomical treatises contain detailed computational procedures for predicting celestial events with impressive accuracy. These procedures functioned as early simulation models formal systems that allowed scholars to project the future behavior of the cosmos.

Simulation requires several key components:

- A model of the system
- Initial conditions
- Rules governing change
- Iterative computation over time

Bhāskara's work included all of these elements. He modeled planetary orbits, established starting positions, defined rules of motion, and calculated future states. This is, in essence, predictive modeling.

Modern AI simulations from climate models to physics engines operate on the same principles, though at far greater speed and scale. The intellectual structure, however, remains strikingly similar.

What makes Bhāskara's simulations particularly noteworthy is their integration of observation and theory. Predictions were continuously compared against empirical data, and discrepancies prompted refinement of the model. This feedback loop is identical to the training-validation cycle used in machine learning today.

## Time, Cycles, and Predictive Intelligence

A defining feature of Indian predictive thought is its deep engagement with time. Unlike linear conceptions of time prevalent in some other traditions, Indian thinkers emphasized cyclicality. Seasons repeat. Planets return. Patterns recur.

Bhāskarāchārya's work reflects this temporal intelligence. His models were designed to capture recurring patterns rather than isolated events. This focus on cycles is essential for long-term forecasting.

Modern predictive analytics increasingly relies on time-series analysis the study of data points indexed in time order. Whether predicting energy demand or disease spread, recognizing cycles and trends is crucial. Bhāskara's astronomical computations can be understood as early forms of time-series modeling, grounded in observation and formalized through mathematics.

By treating time as a structured dimension rather than a random sequence, Bhāskara enabled prediction to become systematic rather than speculative.

## The Roots of Predictive Analytics

Predictive analytics, as practiced today, combines data, mathematical models, and computational power to anticipate future outcomes. While the technology is modern, the intellectual foundations are ancient.

Bhāskarāchārya contributed to these foundations in several key ways:

- He demonstrated that change can be modeled mathematically.
- He showed that future states can be inferred from present conditions.
- He emphasized method over memorization, enabling scalability.
- He integrated validation and correction, recognizing the importance of feedback.

These principles underpin modern AI systems. Machine learning models are trained, evaluated, adjusted, and retrained mirroring the iterative refinement found in Bhāskara's work.

It is important to note that Bhāskara's predictive intelligence was not deterministic in a simplistic sense. He acknowledged uncertainty, approximation, and limitation. Prediction was understood as probabilistic, constrained by available knowledge. This humility is a hallmark of mature intelligence and remains relevant today as AI systems grapple with uncertainty and bias.

# Part II — From Bharat to Baghdad to Boston (The Global Migration of Indian Intelligence)

*Chapter 7*

# The Knowledge Export Nobody Teaches

Modern narratives of scientific progress often follow a deceptively simple arc: ancient Greece gives rise to European enlightenment, which then blossoms into modern science and technology. This story, repeated endlessly in textbooks and popular culture, leaves little room for the civilizations that quietly supplied the intellectual raw materials for that progress. Among the most systematically overlooked contributors is India, whose mathematical, astronomical, and logical systems traveled far beyond its borders shaping the foundations of global science long before Europe's scientific awakening.

This chapter examines one of the least taught yet most consequential processes in intellectual history: the export of Indian knowledge across Persia, the Arab world, and eventually into Europe. This was not a single event but a centuries-long transmission pipeline through which ideas were translated, adapted, renamed, and ultimately absorbed into new civilizational identities. Understanding this process is essential not only for historical accuracy, but for grasping how modern computation, algorithms, and artificial intelligence came to be.

The silence surrounding this knowledge transfer is not accidental. It reflects patterns of power, authorship, and cultural dominance that have shaped how history itself is written.

## India to Persia: The First Intellectual Bridge

Long before the rise of Islam or the establishment of European universities, India and Persia were connected through trade routes, diplomatic exchanges, and scholarly contact. These interactions were not superficial. They involved deep intellectual engagement, particularly in astronomy, mathematics, medicine, and calendrical science.

Persian scholars were drawn to Indian astronomy for practical reasons. Accurate calendars were essential for agriculture, taxation, governance, and religious observance. Indian astronomers had developed sophisticated models for planetary motion, eclipse prediction, and timekeeping models that far exceeded those available in many neighboring regions.

Indian astronomical texts, often referred to generically as Siddhantas, began circulating in Persian intellectual circles as early as the Sassanian period. These works introduced concepts that were revolutionary in their precision and structure, including:

- Systematic trigonometric methods
- Planetary tables based on mathematical models rather than myth
- Computational techniques for large cyclical time periods

What made Indian knowledge particularly attractive was its algorithmic nature. These were not philosophical speculations; they were operational systems. A scholar could apply Indian methods and obtain predictable results. This reliability made them indispensable for Persian administrators and scientists.

The transmission was not one-way imitation. Persian scholars translated, commented upon, and adapted Indian texts to local needs. However, the core mathematical logic remained intact, laying the groundwork for further dissemination westward.

## India to the Arab World: Translation as Transformation

The most decisive phase of India's knowledge export occurred during the early Islamic Golden Age. As the Abbasid Caliphate established itself with Baghdad as its intellectual capital, it embarked on one of the most ambitious translation projects in human history. Scholars were sent across Asia, Africa, and Europe to gather texts that could enhance scientific understanding and administrative efficiency.

India occupied a central place in this effort.

Arabic sources themselves record the import of Indian astronomical and mathematical works, often referred to as Sindhind a linguistic transformation of the Sanskrit word Siddhanta. These texts were translated into Arabic under state patronage, signaling their perceived importance.

Indian mathematics introduced several concepts that would become foundational in the Islamic world:

- The decimal place value system
- The operational use of zero
- Algebraic problem-solving techniques
- Trigonometric functions and tables

What distinguished Indian contributions was not just novelty, but methodological clarity. Calculations were rule-based and generalizable. This allowed Arab scholars to build upon them, extending their scope and applying them to new domains such as navigation, architecture, and finance.

It is crucial to understand that the Arab world did not merely preserve Indian knowledge it expanded it. But expansion requires a foundation.

Without Indian numerical systems, much of Islamic mathematics would have lacked the symbolic power needed for abstraction.

This period marks the transformation of Indian computational thought into a global intellectual currency.

## Indian Texts in Baghdad: The House of Wisdom

At the heart of this transformation stood Baghdad's famed Bayt al-Hikma, or House of Wisdom. Far from being a passive library, it functioned as a state-sponsored research institute, translation bureau, and innovation hub. Scholars fluent in Sanskrit, Persian, Greek, and Arabic worked collaboratively to convert foreign knowledge into usable Arabic texts.

Indian works occupied a privileged position within this ecosystem. Astronomical tables based on Indian models were refined and recalculated. Mathematical treatises incorporating Indian numerals replaced older counting systems that relied on cumbersome symbols or physical aids.

One of the most enduring outcomes of this process was the formalization of algorithmic calculation. Procedures derived from Indian mathematics were rewritten in Arabic prose, making them accessible to a wider scholarly audience.

It is in this context that the term "algorithm" itself emerges derived from the Latinized name of Al-Khwarizmi, a scholar whose work was deeply rooted in Indian numerical systems. While Al-Khwarizmi's contributions were significant, his methods depended fundamentally on Indian place value notation and arithmetic logic.

Baghdad thus became a relay station in the global transmission of Indian intelligence. Ideas were not frozen; they evolved. But evolution does not erase ancestry.

## Sanskrit to Arabic to Latin: The Invisible Pipeline

The next phase of the journey took place as Islamic Spain and Sicily became contact zones between the Arab world and Christian Europe. European scholars, recognizing the advanced state of Arabic science, began translating Arabic texts into Latin from the 12th century onward.

This translation movement is often described as the rebirth of European learning. Yet what is rarely acknowledged is that many of these Arabic texts were themselves translations or adaptations of earlier Indian works.

Thus emerged a three-stage pipeline:

- Sanskrit: Original formulation of mathematical and astronomical systems
- Arabic: Translation, expansion, and systematization
- Latin: Rebranding and institutionalization within Europe

By the time Indian ideas reached Europe, their origins were often obscured by layers of translation and reinterpretation. Indian numerals became "Arabic numerals." Indian algorithms became European arithmetic. Indian astronomy became medieval cosmology.

This was not necessarily a conspiracy, but it was a structural inevitability. Europe encountered these ideas through Arabic intermediaries, and history recorded the immediate source rather than the original one.

The result was a profound misattribution of intellectual heritage one that still shapes modern perceptions of scientific origins.

## Why Origins Were Rebranded

The rebranding of Indian knowledge did not occur in a vacuum. It was shaped by cultural, religious, and political forces that influenced how knowledge was claimed and credited.

Several factors contributed to this process.

First, language barriers played a major role. Sanskrit texts were inaccessible to most European scholars, while Arabic translations were readily available. Credit followed accessibility.

Second, institutional power mattered. European universities, churches, and later colonial administrations controlled curricula and historical narratives. Knowledge that entered through European institutions was labeled accordingly.

Third, colonial ideology retroactively reinforced these omissions. As European empires expanded into Asia, acknowledging India as a source of advanced scientific thought conflicted with the narrative of Western superiority. It was more convenient to portray colonized societies as passive recipients rather than original innovators.

Finally, there was a conceptual bias. Indian knowledge systems often integrated mathematics with philosophy, cosmology, and ethics. Western historiography, shaped by post-Enlightenment categories, struggled to recognize such synthesis as "science," even when the results were mathematically rigorous.

The consequence of these forces was not just historical oversight, but intellectual amnesia.

## The Cost of Forgetting

The suppression or dilution of India's role in global scientific development has had lasting effects. It has shaped educational systems, influenced research priorities, and distorted public understanding of how intelligence evolves.

When the history of algorithms is taught without acknowledging Indian foundations, it reinforces the myth that advanced computation is a purely Western invention. This myth, in turn, affects how societies perceive their own capacity for innovation.

For India, the cost has been particularly high. A civilization that once exported mathematical intelligence to the world came to see itself, under colonial influence, as technologically backward. This psychological inversion remains one of colonialism's most enduring legacies.

Reclaiming this history is not about national pride. It is about restoring continuity to the story of human intelligence.

## Knowledge as Migration, Not Miracle

One of the most important lessons of this chapter is that intelligence does not emerge in isolation. It migrates. Ideas move across borders, adapt to new environments, and acquire new names. Civilizations do not invent knowledge from nothing; they inherit, refine, and transmit it.

India's role in this process was foundational. Its mathematical language provided the syntax through which global computation became possible. Persia served as an early bridge. The Arab world acted as a powerful amplifier. Europe institutionalized the results.

Artificial intelligence, like all complex technologies, sits at the end of this long chain. To understand AI without understanding this migration is to misunderstand intelligence itself.

## Rewriting the Global Intellectual Map

Acknowledging India's contribution does not diminish the achievements of Persian, Arab, or European scholars. On the contrary, it enriches the story by revealing how interconnected human intelligence truly is.

The problem is not that others built upon Indian knowledge. The problem is that the foundation was forgotten.

This chapter does not argue that India alone created global science. It argues that without India, the global scientific trajectory would have been radically different.

The knowledge export nobody teaches is not a footnote. It is a missing chapter in the story of how humanity learned to think in algorithms.

*Chapter 8*

# Al-Khwarizmi & the Birth of the "Algorithm"

Few words in modern science are as ubiquitous and as misunderstood as the word algorithm. It powers search engines, governs financial markets, trains artificial intelligence systems, and shapes human behavior at planetary scale. Yet the story of how this word came into being is rarely told with precision, and even more rarely with honesty. The name "algorithm" is often presented as a purely Islamic or Western mathematical invention, detached from the civilizational soil that made it possible. In reality, the birth of the algorithm is not the story of a single man, but of transmission, translation, and transformation, with Indian mathematics at its core.

Muhammad ibn Musa al-Khwarizmi, the ninth-century scholar whose Latinized name gave rise to the term "algorithm," stands at a crucial crossroads in intellectual history. He did not invent algorithms from nothing. Rather, he systematized, translated, and formalized mathematical methods that had already matured in India centuries earlier. His genius lay not in origination alone, but in codification and portability. He took Indian numerical intelligence and made it mobile across civilizations.

To understand the true birth of the algorithm, one must look beyond the name and examine the architecture of ideas that preceded it.

## Who Coined the Word "Algorithm"?

The word "algorithm" derives from the Latinized form of the name al-Khwarizmi, rendered in medieval texts as Algoritmi. Over time, this proper noun transformed into a common noun, coming to signify a general method or rule-based procedure for calculation.

This linguistic evolution is itself revealing. The word did not initially describe a concept; it described a person. Only later did it come to denote a class of abstract processes. This tells us something important: medieval Europe encountered systematic computation through texts attributed to al-Khwarizmi, and from those texts abstracted the idea of rule-based calculation.

Al-Khwarizmi lived and worked in the House of Wisdom in Baghdad during the Abbasid Caliphate, a period marked by intense translation and synthesis of knowledge from across the known world. Greek philosophy, Persian astronomy, and most crucially Indian mathematics were collected, translated, and studied with unprecedented seriousness.

In one of his most influential works, Kitāb al-ḥisāb al-hindī (The Book of Calculation with Hindu Numerals), al-Khwarizmi presented methods of arithmetic using the Indian decimal place-value system. When this text was translated into Latin in the twelfth century, its opening phrase Dixit Algoritmi ("Thus spoke Algoritmi") became a fixture in European mathematical manuscripts.

Over time, Algoritmi ceased to refer solely to the man and came to signify the method itself. This semantic shift marks a pivotal moment in intellectual history: computation was no longer bound to a specific domain like astronomy or commerce. It became a generalizable process.

However, while the word carries al-Khwarizmi's name, the intellectual content it describes predates him by centuries.

## Algebra from Indian Systems

Al-Khwarizmi is often celebrated as the "father of algebra," largely due to his work Kitāb al-jabr wa'l-muqābala. The term al-jabr would later give rise to the word "algebra." Yet here again, careful historical analysis reveals continuity rather than sudden invention.

Indian mathematicians had been solving algebraic problems long before al-Khwarizmi's time. Quadratic equations, indeterminate equations, and systems of relationships were treated systematically in Sanskrit texts such as those by Aryabhata, Brahmagupta, and Bhāskarāchārya. These works did not merely present solutions; they offered procedural methods step-by-step rules that could be applied repeatedly.

What al-Khwarizmi did was to reframe these methods in a new linguistic and cultural context, stripping them of their poetic sutra-based format and presenting them as prose-based procedures suitable for bureaucratic, legal, and commercial use. This transformation made algebra accessible to administrators, merchants, and engineers, not just scholars.

Crucially, al-Khwarizmi's algebra is algorithmic in nature. It is not concerned with abstract symbolism (which would come later in Europe), but with operational clarity. Problems are stated verbally, and solutions are given as sequences of actions. This is precisely how algorithms function: input, transformation, output.

The Indian origin of these methods is not speculative. Al-Khwarizmi himself acknowledged his sources. His arithmetic text explicitly refers to Indian calculation methods. The structure of his algebraic reasoning mirrors earlier Indian approaches, particularly in its emphasis on balance, reduction, and transformation concepts deeply rooted in Indian mathematical philosophy.

Thus, algebra did not originate as a sudden leap in Baghdad. It was transmitted, adapted, and repackaged. Al-Khwarizmi's contribution lies in making it systematic, teachable, and exportable.

## Decimal & Zero Transmission

Perhaps the most consequential element of al-Khwarizmi's legacy is his role in transmitting the Indian decimal place-value system, including zero, to the rest of the world. Without this transmission, modern computation as we know it would be impossible.

The Indian decimal system is not merely a set of symbols; it is a positional logic system. The value of a digit depends on its position, and zero functions as both placeholder and operator. This system allows for efficient arithmetic, scalability, and abstraction features essential to algorithmic thinking.

Al-Khwarizmi's arithmetic text explained this system in detail, introducing methods for addition, subtraction, multiplication, and division using Indian numerals. These methods were radically simpler and more powerful than Roman numeral arithmetic, which dominated Europe at the time.

When these ideas entered Europe through Latin translations, they encountered resistance. The Church and established institutions viewed the new numerals with suspicion, associating them with foreign influence and potential fraud. Yet their computational superiority could not be denied. Gradually, they replaced older systems.

What is often forgotten is that zero itself was a foreign concept to Europe. Greek mathematics had no true zero. Roman numerals had no placeholder. Without zero, there can be no binary logic, no floating-point computation, no error correction, and no machine intelligence.

Al-Khwarizmi did not invent zero. He served as its vector the medium through which it traveled from Indian abstraction into global infrastructure.

This distinction matters. Transmission is not invention, but it is no less historically significant. Civilizations rise not only by creating knowledge, but by moving it across boundaries.

## The Name That Replaced Its Indian Roots

The final and most sensitive aspect of this story concerns historical attribution. Why did the word "algorithm" come to be associated almost exclusively with al-Khwarizmi, while the Indian origins of algorithmic thinking faded from popular memory?

The answer lies in a convergence of factors: language, power, and historical narrative construction.

Indian mathematical texts were written in Sanskrit, often in compressed sutra form that required oral instruction to decode. This made them difficult to translate without deep cultural immersion. By contrast, Arabic scholars rendered these ideas into expansive prose, making them more accessible to later translators.

When Europe began its intellectual reawakening, it encountered Indian mathematics through Arabic intermediaries, not directly. As a result, attribution followed the most immediate source, not the original origin.

Over time, this intermediate layer hardened into perceived origin. Al-Khwarizmi became the "father of algorithms," just as Arabic numerals became "Arabic," despite their Indian birth.

This process was later reinforced by colonial historiography, which tended to credit Europe and its intellectual ancestors while minimizing non-Western contributions. Indian mathematics was recast as mystical or intuitive, rather than formal and systematic.

The renaming of Indian computation under foreign labels was not merely a semantic shift. It altered how civilizations perceived their own intellectual heritage. India became a source of spiritual wisdom, while the West claimed rational and scientific dominance.

Correcting this narrative does not diminish al-Khwarizmi's brilliance. On the contrary, it places him where he belongs: as a bridge-builder, not an isolated originator.

A clearer lineage would look like this:

- Indian civilization developed formal, rule-based computation
- These methods were transmitted to the Islamic world

- Al-Khwarizmi systematized and taught them
- Europe inherited them through Arabic texts
- Modern computing emerged from this cumulative process

Understanding this lineage restores continuity to the history of intelligence.

## Algorithms as Civilizational Technology

To grasp the deeper meaning of al-Khwarizmi's legacy, one must recognize that algorithms are not merely technical tools. They are civilizational technologies. They encode assumptions about order, predictability, and control.

Indian mathematics approached computation as a harmony between abstraction and reality. Rules existed, but they were embedded in cosmology, ethics, and balance. Islamic scholars emphasized clarity, order, and universality. European thinkers later focused on mechanization and optimization.

Al-Khwarizmi's work sits at the center of this transformation. He took Indian abstraction and gave it bureaucratic precision. He removed cosmological framing and replaced it with operational efficiency. This made algorithms suitable for administration, trade, and governance.

In modern artificial intelligence, this transformation reaches its logical extreme. Algorithms no longer merely assist human decision-making; they increasingly replace it. Understanding their origin reminds us that algorithms were never meant to be value-neutral. They were always expressions of cultural logic.

## From Al-Khwarizmi to Artificial Intelligence

The algorithms that power artificial intelligence today are descendants of the same rule-based thinking al-Khwarizmi helped popularize. Machine learning models rely on:

- Stepwise procedures
- Numerical abstraction

- Error minimization
- Iterative refinement

These are not modern inventions. They are ancient ideas scaled by computation.

What has changed is not the logic, but the speed, scale, and autonomy of execution. Machines now perform in seconds what once took scholars years. But the underlying principles remain unchanged.

Recognizing al-Khwarizmi's role and the Indian foundations beneath it allows us to see artificial intelligence not as an alien force, but as a continuation of humanity's long effort to formalize intelligence.

## Reclaiming the Algorithm's True History

The word "algorithm" carries with it a story of migration, adaptation, and erasure. Restoring its full history is not an exercise in pride, but in accuracy.

India did not invent artificial intelligence in the modern sense. But it invented algorithmic thinking the belief that intelligence can be expressed through rules, symbols, and transformation. Al-Khwarizmi recognized this power, preserved it, and passed it forward.

In doing so, he ensured that Indian intelligence would shape the future even if its name was forgotten.

Today, as algorithms govern economies, influence elections, and shape consciousness itself, understanding their true origin becomes more than academic. It becomes a matter of intellectual responsibility.

The algorithm was never born in isolation.

It was born in India, named in Baghdad, matured in Europe, and now governs the world.

## *Chapter 9*

# How Europe Built on Indian Computation

The rise of European science is often portrayed as a sudden awakening, a dramatic break from the so-called "dark ages" into a period of reason, discovery, and mathematical clarity. Textbooks typically frame this transformation as an internal European phenomenon, driven by Greek rediscovery, Renaissance humanism, and later Enlightenment rationality. Yet this narrative, repeated for centuries, conceals a deeper and more complex reality. Europe did not build its mathematical and computational foundations in isolation. It inherited them. And many of the most essential components of that inheritance originated not in Greece or Rome, but in India.

From number systems and algebraic methods to trigonometry, algorithms, and positional notation, Indian computation formed the invisible backbone of Europe's scientific ascent. These ideas traveled slowly, passing through Persian and Arab scholars, translated into Arabic, and eventually into Latin. Along the way, origins were obscured, names were changed, and credit was reassigned. By the time these concepts became central to European mathematics, their Indian roots were rarely acknowledged.

This chapter examines how Europe built upon Indian computation, beginning with Fibonacci and the Indian number system, moving through Renaissance mathematics, exploring the role of the Church in controlling knowledge, and concluding with the colonial rewriting of scientific history. This is not a polemic against Europe, nor an attempt to deny European contributions. Rather, it is an effort to restore continuity to the historical record and to show that modern Western science is not a purely Western creation, but the product of global intellectual exchange an exchange in which India played a foundational role.

# Fibonacci and the Indian Number System

Leonardo of Pisa, known to history as Fibonacci, occupies a central place in European mathematics. His book Liber Abaci, published in 1202, is often credited with introducing Europe to the Hindu-Arabic numeral system. What is less emphasized is that Fibonacci himself openly acknowledged that this system came from India, transmitted through the Arab world. The numbers he promoted were not European inventions. They were Indian.

Before this transmission, Europe relied largely on Roman numerals, a system poorly suited for computation. Roman numerals lack positional value, making even simple arithmetic cumbersome. Multiplication and division were impractical without counting boards or abacuses. Advanced mathematics was nearly impossible to develop under such constraints. The introduction of the Indian decimal positional system, including zero, changed everything.

The Indian number system offered a radically different approach. It used a base-ten structure, positional notation, and a symbol for zero. This combination allowed numbers to scale infinitely while remaining computationally manageable. It transformed arithmetic from a physical activity into a symbolic one. Calculations could now be performed on paper, abstractly, efficiently, and systematically.

Fibonacci learned this system not in Italy, but through exposure to Arabic mathematics in North Africa. The Arab scholars he encountered had themselves inherited these methods from Indian texts, translated and studied in centers like Baghdad. Fibonacci's contribution was not invention but advocacy. He recognized the superiority of the Indian system and argued for its adoption in Europe.

The resistance he faced is telling. European merchants, clerics, and administrators were suspicious of the new numerals. They appeared foreign, unfamiliar, and even dangerous. Some cities banned their use, fearing fraud. Yet over time, practicality won. The Indian number system spread through trade, accounting, and engineering, gradually replacing Roman numerals.

This adoption laid the foundation for European mathematics. Without positional notation and zero, algebra, calculus, and modern computation would not have been possible. Fibonacci's legacy, therefore, is inseparable from Indian computation. He was a transmitter, not an originator, of a system that had already matured centuries earlier in India.

## Renaissance Mathematics and the Indian Legacy

The European Renaissance is often celebrated as a rebirth of classical knowledge, particularly Greek philosophy and mathematics. While Greek texts did play an important role, this emphasis obscures the extent to which Renaissance mathematics depended on non-European sources, especially Indian computation filtered through the Islamic world.

Algebra, for instance, entered Europe through Arabic texts that drew heavily on Indian mathematical traditions. The very word "algebra" comes from Arabic, but the methods it describes symbolic manipulation, equation solving, and abstraction had deep roots in Indian mathematics. Indian scholars had been solving linear and quadratic equations, working with negative numbers, and exploring indeterminate equations long before these ideas appeared in Europe.

Trigonometry offers another clear example. Sine, cosine, and tangent functions were developed in India as part of astronomical computation. Indian mathematicians constructed detailed trigonometric tables and used them for predictive modeling of celestial motion. These techniques traveled westward, becoming essential tools in European astronomy, navigation, and physics.

During the Renaissance, European mathematicians began to systematize these inherited tools. Figures such as Cardano, Tartaglia, and later Newton and Leibniz built complex theories using numerical and algebraic frameworks that would have been inconceivable without Indian foundations. Yet as these theories took shape within European institutions, their origins were gradually forgotten or deliberately minimized.

The Renaissance was not merely a rediscovery of ancient Greek thought. It was a synthesis. Indian computation, Arabic scholarship, and European intellectual traditions converged to produce new knowledge. To isolate Europe as the sole driver of this transformation is to misunderstand how knowledge actually evolves.

## Church, Science, and Knowledge Control

The role of the Church in shaping European science cannot be ignored. For centuries, the Church functioned as the primary gatekeeper of knowledge in Europe. Monasteries controlled libraries, universities were ecclesiastical institutions, and theological priorities often determined what knowledge was preserved, promoted, or suppressed.

This control had profound implications for how foreign knowledge was received and represented. Ideas that aligned with Christian doctrine or could be framed within it were more easily accepted. Others were resisted or reframed. Indian and Arabic mathematics, stripped of their philosophical and cultural context, were often presented as neutral technical tools rather than products of sophisticated civilizations.

The Church's emphasis on Greek sources over others was not accidental. Greek philosophy had already been integrated into Christian theology through figures like Augustine and Aquinas. Indian thought, by contrast, came from a civilization with entirely different metaphysical assumptions. Acknowledging its influence too openly would have challenged the notion of Europe as the primary heir to rational civilization.

As a result, knowledge transmission was often selective. Techniques were adopted, but origins were obscured. Texts were translated without attribution. Concepts were renamed. Over time, the narrative shifted from one of inheritance to one of independent European development.

This process did not require deliberate malice in every case. It emerged from institutional incentives, cultural biases, and theological constraints. Nevertheless, the outcome was a distorted historical record in which Europe appeared as the primary source of scientific rationality, rather than one participant in a global exchange.

## Colonial Rewriting of Scientific History

The final stage in the erasure of Indian computational influence occurred during the colonial period. As European powers colonized large parts of the world, including India, they did not merely extract resources. They also imposed narratives. History, education, and scholarship were reorganized to reinforce colonial authority.

Colonial education systems presented Europe as the origin of science and reason, while portraying colonized societies as passive recipients or mystical cultures lacking rational thought. Indian mathematics was reduced to isolated curiosities, stripped of its systemic and theoretical depth. Contributions were acknowledged only insofar as they fit within a European-centered framework.

This rewriting of history served a political purpose. By denying colonized peoples intellectual agency, colonial powers justified domination. Science became a marker of superiority, a symbol of European progress contrasted against alleged Eastern stagnation.

Textbooks written during this period continue to shape global education today. Students learn about Greek mathematicians, Renaissance scientists, and Enlightenment thinkers, but rarely about the Indian foundations that made their work possible. When Indian contributions are mentioned, they are often presented as precursors rather than central pillars.

This legacy persists in subtle ways. Even today, discussions of artificial intelligence, computation, and algorithms often begin in Europe or America, ignoring the deeper lineage of ideas. The result is a fragmented understanding of intellectual history, one that obscures the continuity connecting ancient India to modern technology.

*Chapter 10*

# The Suppression of Civilizational Credit

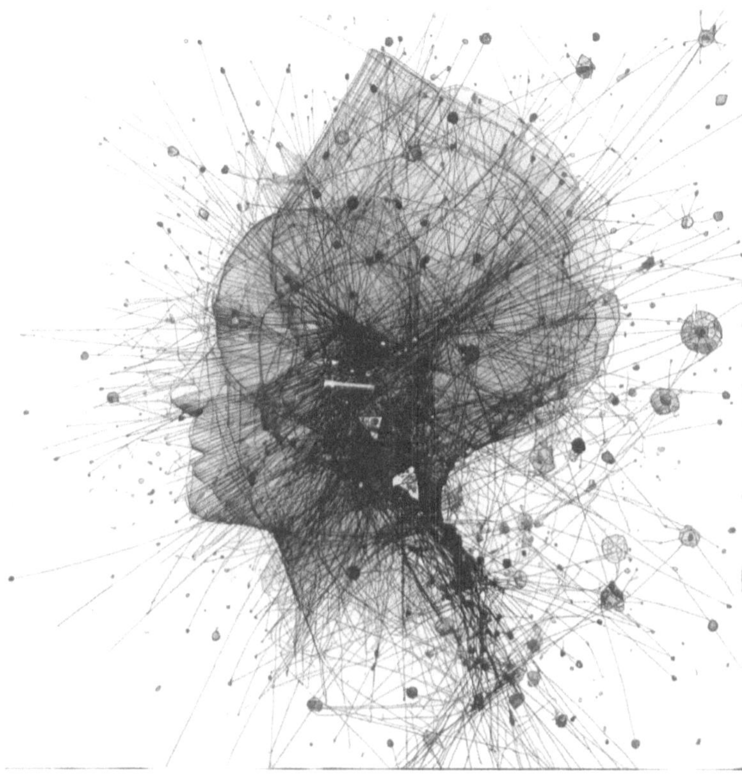

History is not merely a record of events; it is a record of power. What is remembered, what is emphasized, and what is erased are never neutral choices. They reflect the priorities, insecurities, and ambitions of those who control the narrative. In the story of modern science and technology, one of the most consequential acts of power has been the systematic suppression of civilizational credit particularly the minimization, distortion, or outright erasure of non-Western contributions to human knowledge.

This chapter confronts a difficult but necessary truth: the modern global understanding of science, mathematics, and intelligence did not emerge organically from objective scholarship alone. It was shaped often aggressively by colonial education systems, ideological agendas, and political motivations that rebranded inherited knowledge as original Western invention. India's role in the cognitive lineage of mathematics, logic, computation, and prediction was not forgotten by accident; it was obscured by design.

To understand artificial intelligence and modern computation honestly, we must first understand how the historical credit for foundational ideas was reassigned. This chapter examines the mechanisms of that reassignment and the long-term consequences it produced.

## Colonial Education Models: Manufacturing Intellectual Dependency

Colonial rule was not sustained by military force alone. It was maintained through education through the deliberate reshaping of how colonized populations understood knowledge, authority, and themselves. Nowhere was this more evident than in British India.

Colonial education models were explicitly designed to disconnect Indians from their intellectual heritage while positioning Western knowledge as the sole legitimate source of truth. The objective was not merely to educate, but to reorient cognition. A population that believes its past is intellectually barren is far easier to govern than one aware of its civilizational depth.

British education policy in India emphasized English-language instruction, European history, and Western scientific frameworks while systematically marginalizing Sanskrit, Arabic, Persian, and regional knowledge systems. Traditional institutions gurukuls, pathshalas, madrasas, and centers of indigenous learning were either dismantled, defunded, or dismissed as obsolete.

This was not an unintended side effect of modernization. It was openly articulated policy. Colonial administrators argued that indigenous knowledge systems were irrational, unscientific, or inferior, despite relying heavily on knowledge that had itself passed through India centuries earlier.

The result was a manufactured intellectual hierarchy:

- Western knowledge was framed as universal and rational
- Indigenous knowledge was framed as local, mystical, or outdated
- Indian scholars were taught Western conclusions without being taught their Indian origins
- Historical continuity was deliberately severed

This model created a psychological dependency. Indians learned to look outward for validation rather than inward for lineage. Over time, generations were educated into believing that modern science was something imported, not something evolved from their own civilizational foundations.

## The Erasure of Non-Western Knowledge

Erasure does not always occur through destruction. More often, it occurs through omission. When contributions are excluded from curricula, footnotes, bibliographies, and narratives, they fade from collective memory even if their effects remain embedded in modern systems.

Indian mathematical and scientific contributions suffered precisely this fate. Concepts such as zero, decimal notation, algebraic methods, trigonometry, and early ideas of limits and infinity were absorbed into European scholarship through Arabic intermediaries. Yet as these ideas entered Europe, their origins were gradually stripped away.

The process followed a familiar pattern. Indian ideas were translated into Arabic, then into Latin. Over time, attribution shifted from Indian scholars to Arab translators, and eventually to European mathematicians who refined, systematized, or popularized the concepts. Refinement was mistaken for invention. Transmission was mistaken for origin.

By the time modern science textbooks emerged in the nineteenth and twentieth centuries, Indian contributions were often reduced to brief mentions or eliminated altogether. Zero became a "Hindu-Arabic numeral," then simply "Arabic numerals," and finally just "numbers." Algorithms were named after Al-Khwarizmi, with little discussion of the Indian mathematical systems that informed his work.

This erasure was not limited to mathematics. Linguistics, logic, medicine, and astronomy were similarly reframed. Sanskrit grammar, one of the most formal linguistic systems ever devised, was studied in Europe without acknowledging its foundational relevance to modern computational linguistics. Indian astronomy was cited selectively, while its predictive accuracy and methodological rigor were downplayed.

Erasure, in this sense, was cumulative. Each omission reinforced the next, until an entire civilizational contribution was rendered invisible.

## Textbook Whitewashing and the Illusion of Linear Progress

Modern textbooks often present scientific progress as a linear, Western-driven ascent: Ancient Greece, followed by a long dark age, followed by the European Renaissance, Enlightenment, and modern science. This narrative is tidy, comforting, and deeply misleading.

The so-called "dark ages" were not dark everywhere. While Europe experienced political and institutional decline, India, the Islamic world, and China continued to develop sophisticated systems of knowledge. Yet textbooks frequently compress these centuries into a void, implying that intellectual activity simply paused until Europe resumed it.

This whitewashing serves a powerful ideological function. It positions Western civilization as the natural heir to Greek rationality and the inevitable leader of human progress. Non-Western civilizations are cast as passive transmitters or, worse, irrelevant bystanders.

In such narratives:

- Indian mathematicians appear as precursors at best, curiosities at worst
- Contributions are presented without context, lineage, or influence
- Complex systems are reduced to isolated "discoveries"
- The global nature of knowledge evolution is ignored

Textbooks rarely explain how European scholars accessed ancient knowledge, why Arabic translations were central, or how Indian texts informed those translations. Instead, ideas appear suddenly in Europe, seemingly generated in isolation.

This framing has consequences. It shapes how students, researchers, and policymakers understand innovation. It reinforces the belief that creativity and intelligence are geographically or culturally concentrated. And it subtly legitimizes contemporary imbalances in scientific authority and funding.

Whitewashing does not deny that others existed. It denies that they mattered.

## The Politics of Scientific Ownership

Scientific ownership is not merely about who first had an idea. It is about who gets recognized, cited, funded, and remembered. Ownership confers authority, and authority translates into power.

During the colonial era, claiming ownership of scientific progress served multiple political purposes. It justified imperial dominance by portraying colonizers as bearers of enlightenment. It framed colonized societies as beneficiaries rather than contributors. And it established Western institutions as the arbiters of truth.

Even after formal colonialism ended, these structures persisted. Academic journals, universities, and research bodies continued to privilege Western frameworks and references. Non-Western knowledge was often required to be "validated" through Western methodologies before being taken seriously.

This asymmetry extends into the modern age of artificial intelligence. AI research today is dominated by institutions in North America and Europe. The historical narrative supporting this dominance assumes that computation, algorithms, and intelligence are Western inventions, rather than global inheritances.

When ownership is framed this way, it becomes easier to marginalize alternative perspectives on ethics, cognition, and intelligence. Indian philosophical approaches to consciousness, causality, and responsibility are often dismissed as "cultural" rather than foundational, despite addressing questions that AI now struggles to answer.

The politics of ownership thus shape not only history, but the future of technology.

## Psychological Colonization and Internalized Erasure

Perhaps the most damaging consequence of suppressed civilizational credit is not external misrepresentation, but internalized doubt. When a civilization is taught that its greatest contributions lie in spirituality rather than rationality, it begins to accept a false division between intellect and ethics.

In post-colonial India, generations of students were trained to excel in Western frameworks without being taught that their ancestors helped create the foundations of those frameworks. This produced brilliance without confidence technical skill without historical grounding.

Many Indian scientists, engineers, and technologists succeeded globally, yet rarely questioned why their education omitted their own civilizational lineage. The result was a silent assumption that innovation must always be borrowed, never reclaimed.

This psychological colonization is subtle but profound. It discourages original synthesis. It limits imagination. It reinforces dependency on external validation.

Reclaiming civilizational credit is not about nostalgia or nationalism. It is about restoring intellectual continuity. A civilization that understands its past contributions is better equipped to shape its future.

## Suppression vs. Complexity: Avoiding False Extremes

It is important to approach this subject with nuance. Suppression of civilizational credit does not imply a conspiracy in which all Western scholars acted maliciously. Knowledge transmission is complex. Ideas evolve. Contributions overlap.

However, acknowledging complexity does not negate patterns. When omissions consistently disadvantage certain civilizations and advantage others, structural bias must be recognized.

The issue is not that Europe contributed nothing clearly it did. The issue is that Europe was positioned as the sole source of modern rationality, while others were framed as peripheral.

Restoring balance does not diminish Western achievements. It contextualizes them. It situates them within a global lineage rather than an isolated narrative.

## Why This Suppression Matters for Artificial Intelligence

Artificial intelligence is not value-neutral. It encodes assumptions about intelligence, reasoning, language, and ethics. If the historical foundations of these concepts are misrepresented, the systems built upon them will inherit those distortions.

When AI is framed exclusively as a Western technological achievement, alternative models of intelligence are marginalized. Holistic reasoning, ethical integration, and systems thinking hallmarks of Indian intellectual traditions are often treated as philosophical add-ons rather than core design principles.

Recognizing India's role in the cognitive lineage of AI expands the design space. It invites new approaches to language processing, ethical alignment, and human–machine interaction.

In this sense, reclaiming civilizational credit is not about the past alone. It is about ensuring that the future of intelligence is genuinely global.

# Part III — The Birth of Modern Computing

*Chapter 11*

# From Mechanical Logic to Electric Logic

The modern digital world is often presented as a sudden rupture in history: a clean break between the pre-computer age and the era of electronics, software, and artificial intelligence. In this simplified narrative, intelligence leaps from philosophy to physics, from ideas to machines, almost overnight. Yet this portrayal conceals a long and intricate transition one in which human logic was first mechanized, then formalized, and only later electrified. The birth of computing was not a technological accident; it was the culmination of centuries of work to externalize reasoning itself.

This chapter traces that transformation. It follows the path from mechanical logic to electric logic, examining how abstract thought was progressively translated into systems that could operate without human intuition. Central to this journey are figures such as Gottfried Wilhelm Leibniz, George Boole, Charles Babbage, and Ada Lovelace, whose work collectively turned reasoning into machinery. Yet this story is incomplete without a deeper comparison between Indian logic, which sought harmony, cyclicality, and meaning, and industrial logic, which prioritized efficiency, control, and scale.

Understanding this transition is essential for understanding artificial intelligence. AI did not emerge when computers became fast enough; it emerged when logic itself became computable.

## Leibniz and Binary: Reducing Reality to Choice

Gottfried Wilhelm Leibniz stands at a critical crossroads in intellectual history. A philosopher, mathematician, logician, and diplomat, Leibniz was driven by an ambitious dream: to create a universal logical language in which all reasoning could be reduced to calculation. His vision was not merely mathematical but civilizational. He believed that if reasoning could be formalized, disputes could be resolved not through debate or force, but through computation.

At the heart of Leibniz's project was a radical simplification: the idea that all numbers, and by extension all logical propositions, could be represented using just two symbols 1 and 0. Binary arithmetic, which Leibniz formalized in the late seventeenth century, was not initially conceived as a practical engineering tool. It was a philosophical insight into the nature of truth and falsehood, presence and absence, affirmation and negation.

Leibniz saw binary as more than arithmetic. He connected it explicitly to metaphysics and theology, arguing that creation itself could be understood as emerging from nothingness through simple distinctions. This fascination led him to study non-European philosophies, including Chinese and Indian thought. His correspondence reveals a deep interest in systems that used minimal elements to express complex realities.

What Leibniz achieved was not a machine, but a logical substrate. Binary logic reduced complexity to choice. Every statement could, in principle, be broken down into yes/no decisions. This reduction is foundational to digital computing and artificial intelligence, where all data, no matter how rich, is ultimately represented as binary states.

However, Leibniz's binary logic remained largely theoretical in his lifetime. The world lacked the mechanisms to operationalize it. What he provided was a conceptual bridge a way to imagine thought itself as something that could be encoded.

## Boolean Algebra: Logic Becomes Mathematics

If Leibniz provided the philosophical foundation, George Boole provided the formal machinery. In the mid-nineteenth century, Boole took a decisive step by transforming logic from a branch of philosophy into a branch of mathematics. His work demonstrated that logical relationships could be expressed through algebraic equations, manipulated according to strict rules, and solved mechanically.

Boolean algebra introduced a system in which propositions were treated as variables that could take on values such as true or false. Logical operations AND, OR, NOT were defined mathematically, allowing complex chains of reasoning to be reduced to symbolic manipulation.

This shift cannot be overstated. With Boolean algebra, reasoning no longer required understanding or meaning. It required rule compliance. A system could process logical relationships without knowing what those relationships referred to. This was a profound step toward machine intelligence.

Boole's work effectively severed logic from consciousness. Reasoning became an abstract process independent of the human mind. This abstraction made it possible to imagine machines that could "think," not by understanding, but by following rules.

What is particularly striking is how closely Boolean logic mirrors earlier Indian logical traditions, especially those found in Nyāya philosophy, where propositions, negations, and inference chains were analyzed with formal precision. Yet while Indian logic integrated epistemology, ethics, and metaphysics, Boolean logic deliberately stripped reasoning down to its bare operational core.

This narrowing was necessary for industrialization. Machines do not require meaning; they require instructions. Boolean algebra provided those instructions in a form that could later be implemented physically through circuits and switches.

## Charles Babbage: The Engine That Thought in Steps

While Leibniz and Boole transformed logic conceptually, Charles Babbage attempted to give it physical form. Living in the heart of the Industrial Revolution, Babbage was acutely aware of the limitations of human calculation. Errors in mathematical tables had serious consequences for navigation, engineering, and finance. His solution was radical: remove the human calculator entirely.

Babbage designed two revolutionary machines the Difference Engine and the Analytical Engine. The former was intended to compute polynomial functions automatically. The latter, far more ambitious, contained the essential components of a modern computer: a memory store, a processing unit, conditional branching, and programmability.

The Analytical Engine was never completed, largely due to technological and financial constraints. Yet its design reveals a mind thinking unmistakably in algorithmic terms. Babbage envisioned a machine that could execute sequences of operations based on instructions, store intermediate results, and alter its behavior depending on outcomes.

This was mechanical logic in its purest form. Gears and levers replaced neurons and intuition. Reasoning became motion.

Babbage's engines were shaped by industrial logic. Precision, repeatability, and efficiency were paramount. The goal was not insight, but reliability. Unlike earlier mathematical traditions that sought cosmic harmony or philosophical truth, Babbage's work was driven by the needs of an industrial society that demanded scale and control.

Yet despite its limitations, Babbage's vision marked a turning point. Logic was no longer confined to paper or mind. It had entered the realm of machinery.

## Ada Lovelace: The First Programmer and the Limits of Machines

If Babbage designed the hardware of computation, Ada Lovelace grasped its deeper implications. Often described as the world's first computer programmer, Lovelace recognized that the Analytical Engine was not merely a calculating device. It was a symbol-processing machine.

In her famous notes on Babbage's work, Lovelace described how the engine could manipulate symbols according to rules, opening the possibility of composing music, creating art, or processing language provided these domains could be expressed symbolically. This insight places her astonishingly close to modern conceptions of artificial intelligence.

Yet Lovelace also articulated a crucial limitation. She argued that machines could only do what they were instructed to do. They could follow rules, but they could not originate ideas. This observation, later termed "Lady Lovelace's Objection," continues to shape debates about machine creativity and consciousness.

What Lovelace understood, perhaps better than any of her contemporaries, was the difference between mechanical intelligence and human intelligence. Machines could manipulate form, but not meaning. They could execute algorithms, but not possess intention.

Her perspective offers a critical counterbalance to industrial logic. While Babbage sought automation, Lovelace reflected on agency. She saw computation as powerful, but bounded.

In many ways, Lovelace's insight echoes Indian philosophical distinctions between calculation (karma) and awareness (chaitanya). Action can be automated; consciousness cannot. This distinction remains central in contemporary AI ethics.

## From Gears to Circuits: The Arrival of Electric Logic

The transition from mechanical to electric logic occurred in the twentieth century, when Boolean algebra found a physical medium in electrical circuits. Switches could represent binary states. Circuits could implement logical operations. Speed and scalability increased exponentially.

Electric logic solved the problems that mechanical computation could not. It eliminated friction, reduced error, and enabled miniaturization. What had once required rooms of machinery could now fit on a board, and later on a chip.

This transformation completed the journey Leibniz had imagined centuries earlier. Logic was no longer merely formal; it was executable at scale. With electricity, algorithms could operate faster than human thought, paving the way for artificial intelligence.

Yet something was lost in this acceleration. As logic became faster and more efficient, it also became more detached from meaning. Electric logic inherited the industrial emphasis on output and optimization, often at the expense of reflection.

## Indian Logic vs Industrial Logic: Two Visions of Intelligence

To fully understand the implications of this transition, it is necessary to contrast two fundamentally different traditions of reasoning: Indian logic and industrial logic.

Indian logic, as developed in systems such as Nyāya, Sāṃkhya, and Buddhist epistemology, treated reasoning as one aspect of a broader inquiry into reality. Logic was inseparable from ethics, perception, and liberation. Knowledge was evaluated not only by correctness, but by its effect on human understanding and well-being.

Industrial logic, by contrast, emerged from the needs of mechanization. Its priorities were efficiency, scalability, and predictability. Truth was operational, not experiential. A system was correct if it produced the desired output.

These differences can be summarized succinctly:

- Indian logic emphasized context, causality, and consciousness
- Industrial logic emphasized efficiency, control, and abstraction

Neither tradition is inherently superior, but their divergence has profound consequences. Modern artificial intelligence is almost entirely built on industrial logic. It excels at optimization, pattern recognition, and automation. Yet it struggles with meaning, ethics, and wisdom.

This imbalance explains many of the anxieties surrounding AI today. Systems that optimize without understanding can amplify bias, concentrate power, and erode human agency. Indian logic offers not a replacement, but a corrective a reminder that intelligence divorced from conscience is incomplete.

## Mechanical Minds, Electric Speed, Human Questions

The journey from mechanical logic to electric logic represents one of humanity's greatest intellectual achievements. It demonstrates our ability to externalize thought, to build systems that reason faster and more consistently than we can.

Yet this achievement also forces us to confront fundamental questions. If logic can be mechanized, what remains uniquely human? If intelligence can be automated, where does responsibility lie?

The figures explored in this chapter did not merely invent technologies; they reshaped how intelligence itself is understood. Leibniz reduced reasoning to choice. Boole mathematized logic. Babbage mechanized computation. Lovelace humanized its limits.

Together, they prepared the ground for artificial intelligence. But the story does not end with electricity. As AI advances, humanity must decide which logic will guide it the narrow efficiency of industrial systems, or the broader wisdom traditions that once sought harmony between knowledge and meaning.

The future of intelligence may depend not on faster machines, but on remembering the philosophies that shaped thought long before machines existed.

## *Chapter 12*

# The Digital Machine is Born

The birth of the digital machine marks one of the most profound turning points in human history. It was not merely the creation of a new tool, but the externalization of abstract intelligence into physical form. For the first time, logic, calculation, and symbolic reasoning were no longer confined to the human mind or to written notation. They were embedded into machines that could operate autonomously, relentlessly, and at speeds that shattered all previous limits.

Yet this moment did not arrive suddenly, nor did it emerge from a single civilization or mind. The digital machine is the result of a long intellectual lineage stretching from ancient numerical abstraction to modern electronic engineering. Its emergence in the twentieth century represents the materialization of algorithmic thought, a thought system whose foundations were laid centuries earlier.

This chapter traces how computation crossed a critical threshold: from conceptual method to physical reality. From wartime urgency to silicon logic, from vacuum tubes to transistors, from exponential scaling to the enduring supremacy of zero, the digital machine's story is one of convergence of mathematics, physics, logic, and necessity.

# ENIAC and Wartime Computing

The immediate catalyst for the digital machine was not curiosity, but crisis. World War II demanded computational power at a scale no human workforce could provide. Artillery trajectories, bomb blast simulations, cryptographic analysis, radar calculations, and logistical planning all required rapid, repeatable, and error-resistant computation. Mechanical calculators and human "computers" were no longer sufficient.

It was within this context that the Electronic Numerical Integrator and Computer (ENIAC) was born. Completed in 1945 in the United States, ENIAC is often described as the first general-purpose electronic digital computer. It occupied an entire room, weighed over thirty tons, and consumed enormous amounts of electrical power. Yet despite its physical bulk, ENIAC represented a conceptual breakthrough: calculation executed electronically rather than mechanically.

ENIAC did not "think" in the human sense, but it followed explicitly defined procedures at unprecedented speed. It could perform thousands of operations per second, transforming weeks of manual calculation into minutes of machine execution. This speed was not simply a quantitative improvement; it was a qualitative leap. At such speeds, new kinds of problems became solvable.

Importantly, ENIAC was not programmed in the modern sense. Its "programs" were configured by physically rewiring panels and setting switches. This made it inflexible by later standards, but the principle was clear: logic could be encoded into hardware, and hardware could execute logic deterministically.

The conceptual lineage here is unmistakable. ENIAC was an embodiment of algorithmic thinking: a machine that executed step-by-step procedures on numerical symbols. These procedures did not originate in the twentieth century. They were the descendants of mathematical rule systems refined across civilizations.

What changed was not the logic, but the substrate. Algorithms moved from parchment and mind into electrons and circuits.

## From Human Computers to Electronic Intelligence

Before machines like ENIAC, computation was largely performed by teams of human calculators often women who followed written procedures to compute tables, trajectories, and predictions. These humans were called "computers," a term that underscores the continuity between human and machine computation.

Digital machines did not replace intelligence; they replicated and accelerated procedural reasoning. Every step a machine executed was first conceptualized by humans using symbolic logic and mathematics. The digital computer is therefore not a rival to human intelligence, but its technological extension.

This distinction matters deeply when discussing artificial intelligence. The digital machine was not designed to be conscious or creative. It was designed to be precise, repeatable, and obedient to rules. These characteristics reflect a particular vision of intelligence: one grounded in formalism rather than intuition.

This vision aligns closely with ancient computational traditions that emphasized rule-based reasoning. The digital machine represents the moment when this vision became physically irreversible.

## The Transistor Revolution

While ENIAC proved that electronic computation was possible, it also exposed its limitations. Vacuum tubes, the primary switching components used in early computers, were fragile, energy-intensive, and prone to failure. ENIAC contained nearly eighteen thousand vacuum tubes, and even a single failure could disrupt operations. Maintenance was constant, and scalability was limited.

The true birth of the digital age came not with ENIAC, but with the transistor.

Invented in 1947 at Bell Laboratories, the transistor replaced vacuum tubes with a smaller, more reliable, and vastly more efficient switching device. Unlike vacuum tubes, transistors generated little heat, consumed minimal power, and could be miniaturized. This made it possible to build computers that were not room-sized anomalies, but scalable systems.

At a conceptual level, the transistor did something remarkable. It transformed abstract binary logic into a stable physical phenomenon. By controlling the flow of electrons, transistors could represent binary states on and off, one and zero with extraordinary reliability.

This is where ancient abstraction meets modern physics. Binary logic, which underlies all digital computation, depends entirely on the distinction between presence and absence, signal and no signal. This distinction is meaningless without zero. The transistor is, in effect, a physical manifestation of śūnya: controlled nothingness.

The transistor revolution did not merely improve computers; it democratized computation. Machines became smaller, cheaper, and more accessible. Computation moved out of government laboratories and into universities, corporations, and eventually homes.

Each reduction in size expanded the scope of computation. Smaller machines meant more of them. More machines meant more data. More data meant new forms of analysis. The digital machine was no longer a specialized wartime instrument; it was becoming a general-purpose infrastructure.

## Integrated Circuits and the Scaling of Intelligence

The next leap came with the development of integrated circuits, which allowed multiple transistors to be fabricated onto a single piece of silicon. This eliminated the need for individual wiring and enabled unprecedented levels of miniaturization.

Integrated circuits did more than increase speed. They changed the economics of computation. When logic gates could be printed en masse, the cost of computation plummeted. This economic shift is as important as the technological one. Intelligence, once scarce and localized, began to scale.

With integrated circuits, computers became platforms rather than projects. Software emerged as a distinct layer, separate from hardware. Algorithms could now be written once and executed anywhere compatible hardware existed.

This separation echoes earlier civilizational shifts. Just as mathematical rules could be abstracted from specific problems, computation could now be abstracted from specific machines. Intelligence became portable.

## Moore's Law

In 1965, Gordon Moore, co-founder of Intel, observed that the number of transistors on a chip appeared to double approximately every two years. This empirical observation came to be known as Moore's Law, and it would shape the trajectory of digital technology for decades.

Moore's Law is not a physical law. It is an economic and engineering pattern, sustained by relentless innovation and investment. Yet its implications are staggering. Exponential growth means that improvements compound, producing capabilities that quickly exceed human intuition.

As transistors doubled, computers became faster, smaller, and cheaper. Tasks that once required supercomputers could be performed on personal devices. Eventually, devices would become powerful enough to simulate learning, perception, and decision-making capabilities associated with intelligence itself.

Moore's Law transformed computation from a static tool into a self-accelerating force. Each generation of machines enabled the design of the next. This feedback loop mirrors learning systems, where outputs become inputs for further improvement.

From a civilizational perspective, Moore's Law represents the moment when algorithmic intelligence began to scale beyond human pace. This scaling made artificial intelligence not only possible, but inevitable.

## The Illusion of Novelty

It is tempting to view the digital revolution as entirely modern, disconnected from earlier intellectual traditions. But this is an illusion created by speed and visibility. The core principles driving digital machines symbolic representation, rule-based transformation, abstraction, and error correction are ancient.

What is new is not the logic, but the velocity and autonomy of execution. Digital machines execute logic without fatigue, emotion, or distraction. They apply rules exactly as specified, millions or billions of times per second.

This raises an important philosophical point. The digital machine does not introduce a new kind of intelligence. It introduces a new scale of intelligence. It magnifies procedural reasoning until it becomes transformative.

Understanding this continuity prevents us from mystifying machines. Artificial intelligence is not an alien mind. It is accelerated formal reasoning.

## Why Zero Still Rules

At the heart of every digital machine lies a simple truth: zero still rules.

Digital computation is binary. Every operation reduces ultimately to sequences of ones and zeros. These are not numbers in the traditional sense, but states: voltage present or absent, current flowing or stopped.

Zero is not emptiness here. It is a defined state. Without zero, there is no binary distinction. Without binary distinction, there is no digital logic. Every transistor, every logic gate, every memory cell depends on the controlled representation of nothingness.

This is not a metaphor. It is a physical reality. Modern processors perform trillions of operations per second, each operation involving transitions between zero and one. The entire edifice of artificial intelligence neural networks, language models, predictive systems rests on this foundation.

The philosophical implications are profound. Intelligence, when mechanized, depends not on continuous presence, but on discrete absence. Meaning emerges not from fullness alone, but from contrast, from difference, from negation.

This idea was understood conceptually long before it was engineered physically. The Indian conception of śūnya as both void and potential finds its ultimate expression in digital machinery. Zero is not the enemy of intelligence; it is its enabling condition.

## Digital Machines and the Future of Intelligence

The birth of the digital machine did not end with transistors or Moore's Law. It set in motion a transformation that continues today. Artificial intelligence systems now learn patterns, generate language, recognize images, and influence decisions at global scale.

Yet despite their apparent complexity, these systems remain faithful to their origins. They are rule-based, numerical, and grounded in binary logic. They do not escape zero; they depend on it.

As digital machines grow more powerful, the question is no longer whether they can compute, but how they should be governed. The same formalism that gives machines power also strips them of conscience. Ethics, meaning, and responsibility cannot be reduced to zeros and ones without loss.

This is where civilizational memory becomes essential. Understanding where digital machines come from intellectually, mathematically, philosophically allows us to shape where they go.

The digital machine was born in war, matured in commerce, and now shapes culture. It stands at the intersection of ancient abstraction and modern urgency.

It is not the end of intelligence.

It is the latest chapter in humanity's long effort to understand, formalize, and extend the power of the mind one zero at a time.

*Chapter 13*

# The Mathematical DNA of Artificial Intelligence

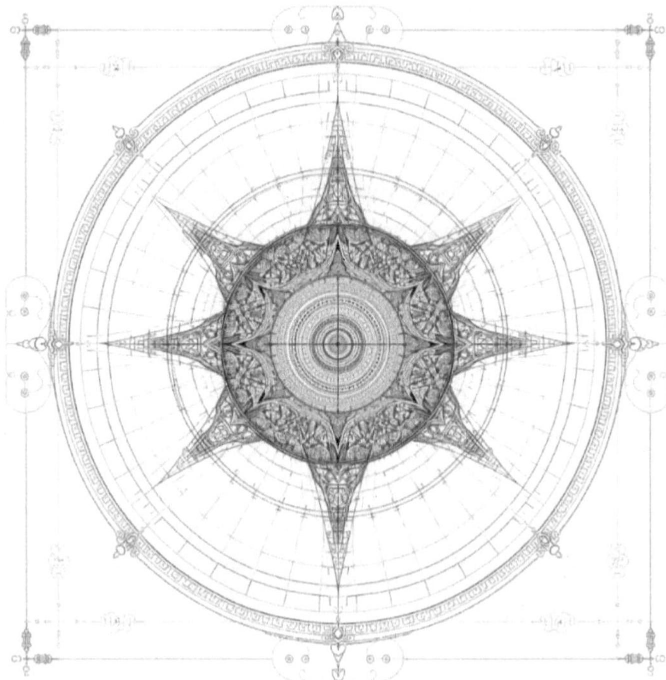

Artificial Intelligence is often spoken of as a technological marvel born from silicon, data centers, and modern code. In popular imagination, AI appears as something fundamentally new, an unprecedented leap that separates the digital age from all that came before it. Yet this perception collapses under careful examination. Artificial Intelligence is not a rupture in human history; it is an extension of it. Beneath the layers of software, hardware, and computational power lies a deeper structure a mathematical foundation that predates modern machines by centuries, even millennia. This foundation can be thought of as the mathematical DNA of AI, the inherited logic that determines how artificial systems learn, reason, and predict.

At its core, AI is not magic. It is mathematics in motion. Every neural network, every recommendation system, every language model is governed by a combination of linear algebra, probability theory, optimization, and graph-based reasoning. These are not incidental tools chosen for convenience; they are structural necessities. Without them, AI could not function. What is rarely acknowledged, however, is that many of the conceptual roots of these mathematical disciplines trace back to ancient Indian thought, where abstraction, formal reasoning, and predictive modeling were developed with remarkable sophistication.

This chapter explores the mathematical DNA of Artificial Intelligence by examining its four essential pillars linear algebra, probability theory, optimization, and graph theory and then tracing their intellectual lineage back to Indian mathematics. The goal is not to claim that ancient Indian scholars built AI systems, but to demonstrate that they developed the cognitive frameworks that make AI possible. Intelligence, whether human or artificial, emerges from structure. India was among the earliest civilizations to formalize that structure.

## Linear Algebra: The Geometry of Intelligence

Linear algebra forms the backbone of modern AI. Every neural network, regardless of its complexity, operates through vectors, matrices, and linear transformations. Inputs are represented numerically, transformed through weighted connections, and combined to produce outputs. This process, repeated across layers, enables machines to recognize patterns, classify information, and generate predictions.

At a technical level, linear algebra allows AI systems to represent high-dimensional data efficiently. Images, text, sound, and sensory inputs are converted into numerical vectors. These vectors are multiplied by matrices of weights, summed, and passed through activation functions. Learning occurs when these matrices are adjusted to minimize error. Without linear algebra, none of this would be possible.

Yet the idea of representing complex phenomena through structured numerical relationships did not originate in the modern era. Ancient Indian mathematics demonstrated a deep understanding of numerical abstraction and geometric reasoning. Indian scholars treated numbers not merely as counting tools, but as entities capable of expressing relationships, proportions, and transformations.

Indian trigonometry, developed for astronomical purposes, exemplifies this mindset. Functions equivalent to sine and cosine were used to model cyclical motion and spatial relationships. These functions were tabulated, refined, and applied to predictive calculations. At a conceptual level, this is linear algebraic thinking: representing physical reality through mathematical mappings.

Moreover, Indian mathematicians were comfortable working with multidimensional relationships long before formal vector spaces were defined in Europe. Their astronomical models involved multiple variables time, position, angle, and velocity combined through systematic calculation. This approach mirrors how AI systems today handle multi-dimensional data spaces.

What matters here is not the notation, but the abstraction. Linear algebra is fundamentally about structure how components relate, transform, and combine. Indian mathematics cultivated this structural thinking early, embedding it into disciplines like astronomy, architecture, and music. Artificial Intelligence inherits this way of thinking, even if its modern practitioners rarely acknowledge its origins.

## Probability Theory: Reasoning Under Uncertainty

If linear algebra gives AI structure, probability theory gives it flexibility. Modern AI does not operate in a world of certainty. It deals with noisy data, incomplete information, and unpredictable environments. Probability allows AI systems to reason under uncertainty, to estimate likelihoods, and to make informed decisions without absolute knowledge.

In machine learning, probability governs classification, prediction, and inference. Bayesian models update beliefs based on new evidence. Neural networks output probability distributions rather than definitive answers.

Language models estimate the likelihood of one word following another. Everywhere in AI, probability serves as the bridge between data and decision.

The roots of probabilistic thinking are often traced to early modern Europe, particularly to gambling problems and statistical analysis. While formal probability theory as a named discipline emerged later, the conceptual foundations of probabilistic reasoning existed much earlier in Indian thought.

Indian philosophy and mathematics both grappled deeply with uncertainty, chance, and causality. Concepts such as karma were not deterministic in a simplistic sense; they involved probabilistic relationships between actions and outcomes across time. This worldview encouraged thinking in terms of tendencies, likelihoods, and conditional effects rather than fixed outcomes.

In mathematical contexts, Indian scholars worked extensively with combinatorics and permutations, particularly in the study of poetic meters, prosody, and logic. These studies required systematic enumeration of possibilities, an essential component of probability theory. The famous work on combinatorics attributed to Pingala, for example, demonstrates binary enumeration centuries before similar ideas appeared in Europe.

Indian astronomers also incorporated uncertainty into their predictive models. Observational error was acknowledged, approximations were refined, and predictions were continually adjusted. This iterative refinement mirrors the probabilistic updating processes used in modern AI systems.

The key insight is that probability is not merely a mathematical technique; it is a mode of reasoning. It reflects an acceptance that knowledge is partial and that intelligence involves managing uncertainty. This philosophical orientation was deeply embedded in Indian intellectual traditions, long before probability equations were formalized in Western notation.

Artificial Intelligence inherits this orientation. When an AI system predicts, it does not claim certainty; it assigns confidence. This probabilistic mindset is ancient, even if its equations are modern.

## Optimization: Learning as Error Reduction

Optimization lies at the heart of learning. In Artificial Intelligence, learning is defined as the process of minimizing error. A model makes predictions, compares them to reality, calculates a loss, and adjusts itself to reduce that loss. This process repeats until the system reaches an optimal or near-optimal state.

Gradient descent, one of the most fundamental algorithms in machine learning, embodies this principle. It involves moving step by step in the direction that reduces error most efficiently. This idea iterative improvement through incremental adjustment is central to AI.

Optimization is not limited to machine learning. It appears in operations research, decision theory, and control systems. AI constantly balances competing objectives: accuracy versus speed, exploration versus exploitation, precision versus generalization. Optimization provides the mathematical framework for navigating these trade-offs.

Indian mathematics displayed a sophisticated understanding of optimization long before the term existed. Indian scholars were deeply concerned with efficiency, approximation, and refinement. Their numerical methods often involved iterative procedures that converged toward better solutions.

Bhāskarāchārya's work, for instance, demonstrates an intuitive grasp of limit processes and error minimization. Approximate solutions were improved through successive refinements, guided by mathematical insight rather than brute force. This is optimization in spirit, if not in modern notation.

Indian architectural and engineering traditions also reflect optimization principles. Structures were designed to maximize stability while minimizing material. Astronomical instruments were calibrated for precision within practical constraints. These activities required balancing multiple variables to achieve optimal outcomes.

Optimization, at its core, is about choice. It asks how to act given constraints. Indian philosophical traditions treated choice as a central problem of human existence. The idea that actions should align with an optimal path, guided by knowledge and discipline, resonates strongly with optimization theory.

Artificial Intelligence operationalizes this idea mathematically. It transforms the ancient quest for optimal action into algorithms and loss functions. The continuity between these worlds is intellectual, not mechanical.

## Graph Theory: Networks of Intelligence

Graph theory provides the language for representing relationships. In AI, graphs are everywhere. Neural networks are graphs of connected nodes. Knowledge graphs represent entities and their relationships. Social networks, recommendation systems, and routing algorithms all rely on graph-based models.

Graphs allow AI systems to reason not just about individual elements, but about connections. Intelligence emerges not from isolated data points, but from patterns of interaction. Graph theory makes these patterns explicit.

Although graph theory was formalized as a mathematical discipline in the modern era, the underlying intuition is ancient. Indian thought has long emphasized relationality. Concepts were rarely treated in isolation; they were understood through their connections to other concepts.

Indian logic, grammar, and philosophy all employed network-like reasoning. Paninian grammar, for example, operates as a highly interconnected system of rules, where the application of one rule affects others. This interdependence mirrors the structure of a graph.

Indian cosmology also embraced network thinking. The universe was seen as an interconnected whole, governed by relationships rather than isolated forces. This worldview encourages modeling reality as a system of nodes and links rather than independent entities.

In practical terms, Indian scholars used tabular and relational representations to manage complex information. Astronomical tables, genealogies, and philosophical classifications all reflect graph-like thinking.

Graph theory in AI allows machines to navigate complexity. It enables reasoning over relationships, paths, and influence. While the formal language is modern, the intuition behind it understanding intelligence as relational has deep roots in Indian intellectual traditions.

## How These All Trace Back to Indian Mathematics?

The mathematical DNA of Artificial Intelligence is not a single discovery, but a constellation of ideas. Linear algebra provides structure, probability manages uncertainty, optimization drives learning, and graph theory models relationships. Together, they form the skeleton of intelligent systems.

What unites these disciplines is a shared commitment to abstraction, structure, and formal reasoning. These commitments were hallmarks of Indian mathematics. Indian scholars were less concerned with isolated results than with systems frameworks that could generate knowledge reliably and repeatedly.

Indian mathematics emphasized:

- Symbolic representation
- Rule-based transformation
- Iterative refinement
- Predictive modeling
- Conceptual generalization

These principles are exactly what AI requires. Artificial Intelligence is not about intelligence in a mystical sense; it is about modeling patterns and making decisions within structured frameworks. Indian mathematics excelled at creating such frameworks.

The transmission of these ideas to the modern world was indirect. They traveled through Arabic scholarship, were absorbed into European mathematics, and were eventually formalized in modern notation. Along the way, their origins were often obscured. Yet the intellectual lineage remains.

To acknowledge this lineage is not to diminish modern AI research. It is to deepen our understanding of it. Artificial Intelligence represents the latest chapter in a long story the story of humanity's attempt to formalize intelligence itself.

# Part IV — The Rise of Artificial Intelligence

# Chapter 14

# What Is Artificial Intelligence Really?

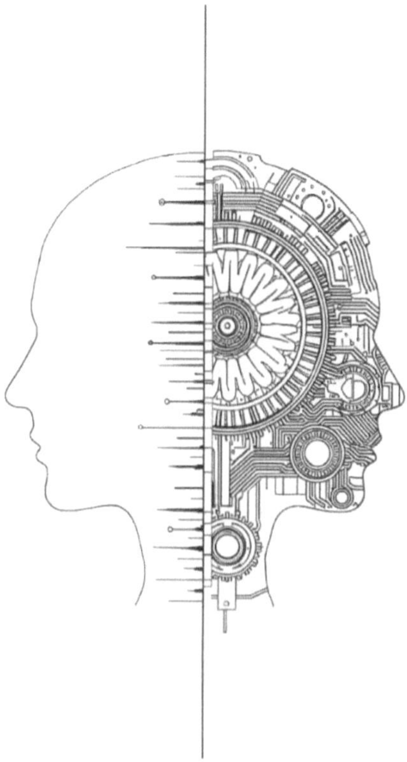

Artificial Intelligence is one of the most frequently used terms of the modern age and one of the least understood. It appears everywhere: in policy documents, marketing slogans, research papers, news headlines, and public debates. Governments claim to regulate it, corporations promise to monetize it, and societies oscillate between celebrating it and fearing it. Yet beneath this noise lies a fundamental problem: there is no shared understanding of what artificial intelligence actually is.

This confusion is not accidental. It arises because artificial intelligence sits at the intersection of technology, philosophy, psychology, mathematics, and power. It is simultaneously a scientific pursuit, an economic engine, and a cultural symbol. As a result, definitions are often shaped less by truth than by convenience. What AI is said to be often depends on who is speaking, what they are selling, or what they fear.

This chapter cuts through the abstraction. It asks a deceptively simple question: What is artificial intelligence, really? To answer it honestly, we must distinguish between definitions and reality, between narrow capability and general intelligence, between machine computation and human cognition. Most importantly, we must confront an uncomfortable truth there are domains of intelligence where ancient human systems still outperform machines.

## Definitions vs Reality

Artificial intelligence is commonly defined as the ability of machines to perform tasks that typically require human intelligence. On the surface, this definition seems reasonable. Yet it collapses under scrutiny. What qualifies as "human intelligence"? And what happens when a task no longer feels intelligent once a machine performs it?

Historically, intelligence has been a moving target. Chess was once considered a pinnacle of human intellect. When machines mastered chess, it was reclassified as brute-force computation. The same happened with arithmetic, pattern recognition, and language translation. Each time machines succeed, the definition of intelligence retreats.

This phenomenon reveals an uncomfortable pattern: intelligence is often defined not by intrinsic properties, but by exclusivity. As long as humans alone can do something, it is called intelligence. Once machines do it, it becomes "just computation."

In reality, artificial intelligence is not intelligence in the human sense. It is artificial inference the execution of statistical, logical, or mathematical procedures that approximate certain outcomes associated with intelligent behavior. Machines do not understand meaning; they model correlations. They do not possess awareness; they optimize objectives.

The gap between definition and reality widens further when commercial incentives enter the picture. Companies label increasingly complex automation as "AI" because the term carries prestige and value. This inflates expectations and fuels fear, while obscuring technical limitations.

To understand AI clearly, we must strip away marketing and myth. Artificial intelligence is not a mind. It is a tool a powerful one but a tool nonetheless. Confusing tools with thinkers is not progress; it is conceptual error.

## The Myth of Autonomous Intelligence

One of the most persistent myths surrounding AI is that it is autonomous in a meaningful sense. Machines are often described as "learning," "deciding," and "thinking" as though these processes arise independently. In truth, every AI system operates within boundaries defined by humans.

Humans choose the data. Humans define the objectives. Humans design the architectures. Humans decide what success looks like. Even systems that appear self-learning are constrained by initial conditions and optimization goals set externally.

Learning, in machines, is not curiosity-driven exploration. It is parameter adjustment. A machine does not ask why; it adjusts weights to reduce error. This distinction is crucial. Intelligence without curiosity, purpose, or self-reflection is fundamentally different from human cognition.

Ancient philosophical traditions recognized this difference intuitively. Intelligence was not equated with output alone, but with intention, awareness, and ethical orientation. A system that produces correct answers without understanding was never considered intelligent it was considered mechanical.

Modern AI excels at execution, not comprehension. It can outperform humans in narrowly defined domains while remaining completely unaware of what it is doing. This is not a flaw; it is a design feature. But mistaking execution for intelligence leads to inflated claims and misplaced fear.

## Narrow AI vs General AI

The distinction between narrow AI and general AI is central to understanding what AI can and cannot do. Narrow AI refers to systems designed to perform specific tasks within well-defined boundaries. These systems dominate today's AI landscape.

Examples include image recognition systems, language translation models, recommendation algorithms, and predictive analytics tools. They perform impressively but only within their training domain. Remove the context, and their intelligence collapses.

General AI, by contrast, refers to a hypothetical system capable of understanding, learning, and applying knowledge across domains in a manner comparable to humans. Such a system would exhibit adaptability, contextual reasoning, and self-directed learning.

Despite frequent claims, general AI does not exist today. What exists are increasingly powerful narrow systems that simulate aspects of intelligence without possessing its underlying structure.

The danger lies in conflating the two. When narrow AI is mistaken for general intelligence, expectations escalate unrealistically. This fuels both utopian fantasies and dystopian fears.

Ancient Indian thought offers a useful lens here. Intelligence was never viewed as task-specific alone. It was holistic integrating cognition, emotion, ethics, and purpose. A system capable of calculation but incapable of judgment was not considered intelligent.

From this perspective, modern AI remains narrow not only technically, but philosophically.

## Machine Intelligence vs Human Intelligence

At a functional level, machine intelligence and human intelligence operate on fundamentally different principles. Machines process symbols. Humans generate meaning. Machines optimize objectives. Humans reflect on values.

Machine intelligence is externally driven. It depends on data provided by humans and feedback mechanisms designed by humans. Human intelligence is internally motivated. It asks questions even when no reward is defined.

Human cognition integrates memory, emotion, intuition, embodiment, and social awareness. These are not add-ons; they are core components of intelligence. Machines, by contrast, operate without subjective experience. They do not feel uncertainty, regret, or insight.

This difference becomes especially clear in ambiguous situations. Humans can reason with incomplete information, conflicting values, and moral dilemmas. Machines struggle in such contexts unless explicitly programmed or constrained.

Ancient systems of intelligence recognized ambiguity as intrinsic to reality. Indian philosophy, for example, embraced paradox, uncertainty, and contextual truth. Intelligence was not about eliminating ambiguity, but navigating it wisely.

Modern AI, in contrast, seeks optimization. It reduces complexity to objective functions. While this is effective for many tasks, it fails when reality resists quantification.

## Consciousness: The Missing Dimension

No discussion of artificial intelligence is complete without addressing consciousness. Machines do not possess it. This is not a temporary limitation; it is a categorical difference.

Consciousness involves subjective experience what philosophers call qualia. It includes awareness of self, perception of meaning, and the capacity for reflection. No AI system today exhibits these traits.

Attempts to equate consciousness with computation misunderstand both. Consciousness is not merely information processing. It is embodied, experiential, and context-sensitive.

Ancient Indian philosophy engaged deeply with consciousness as a primary reality rather than a byproduct. Intelligence was understood as layered, with consciousness at its core. Knowledge without awareness was considered incomplete.

From this standpoint, artificial intelligence can never fully replicate human intelligence. It can mimic behavior, but it cannot experience being.

This realization is not a limitation it is a boundary. Recognizing boundaries allows responsible design. Ignoring them invites illusion.

## Where Ancient Intelligence Still Beats Machines

Despite extraordinary advances, there are domains where ancient human intelligence particularly as developed in long-standing civilizations still outperforms machines.

These include:

- Ethical reasoning, where values conflict and outcomes are uncertain
- Contextual understanding, especially across cultural and historical dimensions
- Wisdom, defined as knowledge applied with restraint and foresight
- Meaning-making, where facts must be integrated into coherent narratives
- Purpose-driven action, where goals arise internally rather than being assigned

Ancient systems trained intelligence holistically. Mathematics was connected to astronomy, astronomy to timekeeping, timekeeping to ethics, and ethics to social harmony. Intelligence was not isolated it was embedded.

Machines, by contrast, excel in fragmentation. They divide problems into discrete units and optimize locally. This makes them powerful, but also blind to the larger picture.

In fields such as governance, justice, education, and moral decision-making, ancient frameworks often provide more robust guidance than algorithmic optimization. This does not mean rejecting technology. It means recognizing that intelligence is more than efficiency.

## The Danger of Overestimating AI

Overestimating artificial intelligence carries serious risks. When machines are assumed to be intelligent in a human sense, responsibility is displaced. Decisions are blamed on algorithms. Accountability dissolves.

This has real-world consequences. Algorithmic bias, surveillance overreach, and automated injustice are often justified as technical inevitabilities rather than human choices embedded in code.

Ancient ethical systems emphasized responsibility. Tools were never moral agents; humans were. Delegating moral authority to machines violates this principle.

AI must remain accountable to human values. Not because machines are evil but because they are indifferent.

## Redefining Intelligence for the Future

If artificial intelligence forces us to rethink intelligence, that is not a weakness it is an opportunity. The rise of AI exposes how shallow many modern definitions of intelligence have become.

Ancient civilizations understood intelligence as a balance between capability, awareness, and ethics. Modern society often reduces it to speed and scale.

A more accurate definition of intelligence would include:

- The ability to model reality
- The capacity to adapt across contexts
- Awareness of consequences
- Ethical orientation
- Purpose beyond optimization

By this measure, AI is powerful but incomplete.

*Chapter 15*

# Neural Networks & the Ancient Brain Model

Few developments in modern artificial intelligence have generated as much fascination, hype, and misunderstanding as neural networks. From facial recognition and speech translation to autonomous vehicles and medical diagnosis, neural networks sit at the heart of today's most powerful AI systems. They are often described as "brain-inspired," a phrase that simultaneously explains their appeal and obscures their limitations. To understand what neural networks truly are and what they are not we must examine both their mathematical foundations and the much older human attempts to model intelligence, perception, and consciousness.

This chapter explores neural networks not merely as a modern engineering achievement, but as part of a long intellectual continuum. It traces the rise of perceptrons and backpropagation, examines the logic of parallel processing, and places these developments in dialogue with ancient Indian models of cognition found in Vedic, Upanishadic, and later philosophical traditions. The aim is not to claim that ancient thinkers invented neural networks, but to show that the conceptual modeling of intelligence long predates machines, and that modern AI captures only a narrow slice of what the human mind has historically been understood to be.

## Perceptrons: The Simplest Artificial Neuron

The perceptron is often introduced as the basic building block of neural networks, a mathematical abstraction inspired loosely by biological neurons. First formalized in the mid-twentieth century, the perceptron models how inputs are combined, weighted, and transformed into an output. At its core, it is a function that takes multiple inputs, multiplies each by a weight, sums the results, and applies a threshold or activation rule to decide the output.

This simple mechanism marked a crucial turning point in the history of artificial intelligence. For the first time, computation was not framed purely as symbolic logic or rule-following, but as pattern recognition. Instead of explicitly programming rules, researchers could allow systems to learn decision boundaries from data.

Yet the perceptron's power lies not in its complexity, but in its repeatability and scalability. A single perceptron is limited it can only solve linearly separable problems but when many perceptrons are combined into layers, the system begins to approximate far more complex functions.

The perceptron represents an important philosophical shift. Intelligence was no longer seen as something that required explicit reasoning. Instead, it could emerge from the accumulation of small, local decisions operating under simple rules. This echoes a long-standing insight in Indian philosophical systems: that complex cognition does not arise from a single commanding entity, but from the interaction of multiple faculties (indriyas, gunas, and vrittis).

However, it is important to be precise. A perceptron is not a neuron. Biological neurons are vastly more complex, influenced by chemical signaling, temporal dynamics, and embodied experience. The perceptron is not a model of the brain; it is a computational metaphor, designed for mathematical convenience rather than biological accuracy.

This distinction matters. When neural networks are described as "thinking like humans," the analogy is often stretched beyond its limits. What perceptrons truly offer is a way to approximate functions, not a window into consciousness.

## Backpropagation: Learning Through Error Correction

If the perceptron provided the basic unit of neural computation, backpropagation provided the method through which networks could learn. Backpropagation is an algorithm that adjusts the weights of a neural network by propagating error backward from the output layer to earlier layers. By calculating how much each weight contributed to the final error, the system incrementally improves its performance.

Conceptually, backpropagation represents a powerful idea: learning as correction, rather than instruction. Instead of being told what to do, the system is told how wrong it is and allowed to adjust itself accordingly. This mirrors human learning in important ways. We do not acquire skills by memorizing perfect rules, but by making mistakes, receiving feedback, and refining our responses.

In Indian epistemological traditions, particularly in Nyāya and Buddhist thought, knowledge acquisition is similarly iterative. Understanding arises through cycles of perception, inference, error, and correction. Truth is not grasped instantaneously; it is approached through refinement. Backpropagation, at a purely mathematical level, reflects this logic.

Yet backpropagation is also deeply limited. It requires a clearly defined objective function, explicit error signals, and large amounts of labeled data. Human learning, by contrast, often occurs without explicit feedback, driven by curiosity, intention, and meaning.

This difference reveals a fundamental gap between artificial learning and human cognition. Neural networks optimize. Humans interpret. Backpropagation is extraordinarily effective at reducing error, but it has no concept of why an outcome matters.

From a philosophical perspective, backpropagation embodies a purely instrumental view of intelligence. Success is defined externally, not internally. The system does not care whether it recognizes a face, diagnoses a disease, or generates a poem. It simply minimizes loss.

This is both the strength and the danger of modern AI. Optimization without understanding can produce remarkable results and equally remarkable failures.

## Parallel Processing: Many Paths, One Outcome

One of the defining features of neural networks is parallel processing. Unlike traditional sequential algorithms, neural networks process information across many nodes simultaneously. Each neuron performs a small computation, and intelligence emerges from their collective interaction.

This idea resonates strongly with ancient Indian models of the mind. In Vedic and post-Vedic thought, cognition is not centralized in a single controlling unit. Instead, it arises from the interaction of multiple components: sensory inputs, mental formations, memory impressions (samskaras), and higher-order awareness (buddhi).

The brain, in this view, is not a command center but a field of activity. Perception, emotion, memory, and reasoning operate in parallel, influencing one another continuously. No single element "thinks" alone.

Neural networks mirror this architecture superficially. Each node performs a simple operation, yet the network as a whole exhibits complex behavior. Pattern recognition, generalization, and robustness emerge not because any individual unit is intelligent, but because the system is distributed.

However, the resemblance has limits. Parallel processing in neural networks is purely mathematical. It lacks the qualitative dimensions present in human cognition emotion, intentionality, and subjective experience. The ancient Indian brain model was never merely computational. It was experiential.

Vedic thought distinguished between manas (sensory-mind), buddhi (intellect), ahamkara (ego-sense), and chitta (storehouse of impressions). These were not isolated modules but interacting processes. Intelligence was understood as layered, contextual, and deeply connected to consciousness.

Modern AI captures only the lowest level of this hierarchy: pattern processing. It does not possess buddhi in the philosophical sense, because it lacks discernment rooted in awareness. It does not possess chitta, because it has no lived memory. It does not possess ahamkara, because it has no sense of self.

Recognizing this difference is essential to avoiding category errors in AI discourse.

## Ancient Brain Models: Cognition Without Mechanism

Ancient Indian thinkers did not dissect brains or build machines. Yet they developed sophisticated models of cognition through introspection, meditation, debate, and logical analysis. These models were not anatomical but functional. They sought to explain how experience arises, how perception is organized, and how error and illusion occur.

In these systems, intelligence was not synonymous with computation. It was a relationship between awareness and experience. The mind was seen as both a tool and an obstacle capable of extraordinary insight, yet prone to distortion.

This contrasts sharply with the modern computational model, which treats intelligence as an information-processing problem. Neural networks do not ask whether their outputs are meaningful. They ask only whether they are statistically effective.

The ancient brain model placed ethical and existential constraints on cognition. Knowledge was valuable only if it reduced suffering, clarified perception, or led to liberation. This is fundamentally different from the industrial logic that governs AI development today.

That difference explains why neural networks, despite their power, often feel alien. They are intelligent without wisdom, capable without comprehension.

## Consciousness Versus Computation

The most important distinction explored in this chapter is the difference between consciousness and computation. Neural networks compute. They transform inputs into outputs through mathematical operations. Consciousness, however, is not an operation. It is an experience.

No neural network, regardless of size or complexity, has subjective awareness. It does not feel, intend, or know. It processes. This is not a failure of engineering; it is a category boundary.

Indian philosophy has long insisted on this boundary. Consciousness (chaitanya or purusha) is not produced by mental activity; it is the ground upon which mental activity appears. The brain does not generate awareness; it organizes experience within awareness.

Modern neuroscience and AI often invert this relationship, treating consciousness as an emergent property of sufficiently complex computation. Yet this claim remains unproven and deeply contested.

Neural networks demonstrate that many cognitive tasks recognition, prediction, classification do not require consciousness. This is an important insight. It suggests that much of what we associate with intelligence is, in fact, mechanical.

But it also reveals what machines cannot do. They cannot care. They cannot suffer. They cannot understand meaning in the existential sense.

This distinction matters enormously as AI systems become embedded in decision-making processes that affect human lives. When machines optimize without awareness, ethical responsibility must remain human.

## The Illusion of the Artificial Brain

The popular metaphor of the "artificial brain" is misleading. Neural networks do not replicate the brain; they replicate certain statistical properties of learning. The brain is embodied, emotional, conscious, and historically situated. Neural networks are abstract, disembodied, and ahistorical.

This does not make them weak. It makes them specialized.

Confusing specialization with general intelligence leads to unrealistic expectations and misplaced fears. Neural networks are powerful tools, but they are not minds.

Ancient cognitive models remind us that intelligence without consciousness is incomplete. Computation without awareness can simulate behavior, but it cannot replace human judgment.

## What Neural Networks Truly Represent

Neural networks represent a triumph of mathematical abstraction, not a replication of human intelligence. They show how far pattern-based learning can go when combined with massive data and computation.

They also expose the limits of a purely industrial conception of intelligence. Optimization alone cannot account for values, meaning, or purpose.

From an Indian philosophical perspective, neural networks correspond to lower cognitive functions pattern recognition without awareness. They operate within prakriti, not purusha. They are mechanisms, not witnesses.

This perspective does not diminish AI; it contextualizes it.

## Toward a Balanced Understanding of Intelligence

The future of artificial intelligence depends on whether humanity recognizes this distinction. Neural networks will continue to grow more capable. They will outperform humans in many tasks. But they will not replace consciousness.

Ancient brain models offer a corrective lens. They remind us that intelligence is not merely the ability to compute, but the ability to understand, choose, and take responsibility.

As AI systems become more embedded in society, the challenge is not to make machines conscious, but to ensure that conscious beings remain accountable for how machines are used.

The ancient brain model does not compete with neural networks. It completes the picture they leave unfinished.

# Chapter 16

# Language Models & Paninian Grammar

Language is the most complex system humanity ever created, and for most of history it resisted formalization. It flowed, evolved, bent rules, and defied strict logic. Philosophers speculated about it, grammarians described it, and poets mastered it, but no civilization before modern times succeeded in turning language into something that could be computed at least not at scale.

Or so the modern story goes.

In the twenty-first century, Large Language Models (LLMs) astonished the world by generating human-like text, translating languages, answering questions, and even reasoning across domains. To many observers, this seemed like a radical rupture with the past: machines suddenly "understanding" language. Yet beneath the apparent novelty lies a deeper continuity. The architecture of modern language models rests on principles that were articulated, with extraordinary precision, more than two thousand years ago by the Indian grammarian Pāṇini.

This chapter explores a claim that initially sounds provocative but becomes increasingly difficult to dismiss: modern AI language models operate in a way that is structurally Paninian. They do not merely resemble ancient grammar metaphorically; they echo its logic at a foundational level. To understand this is not to diminish modern AI, but to situate it within a much older lineage of formal linguistic intelligence.

## Language as a Computational Problem

For a machine, language is not meaning; it is structure. Words become symbols. Sentences become sequences. Meaning is inferred statistically, not intuitively. This is the uncomfortable truth beneath the glamour of artificial intelligence: machines do not "understand" language as humans do. They model it.

To model language, it must be broken down into:

- Discrete units
- Formal relationships
- Transformational rules
- Predictable patterns

This is exactly what Pāṇini achieved in the Aṣṭādhyāyī, composed around the 5th century BCE. Long before computers, he treated language as a formal system, governed by rules that operate on symbols to generate valid expressions.

Modern linguistics often credits Noam Chomsky with introducing the idea of generative grammar. Yet Pāṇini had already constructed a generative grammar of astonishing sophistication one that could produce every valid Sanskrit utterance from a finite set of rules.

This is not poetic exaggeration. It is a technical fact.

## LLM Architecture Explained

To understand the Paninian resonance in modern AI, we must first understand at a conceptual level how Large Language Models work.

An LLM is trained on vast amounts of text. It does not memorize sentences; it learns probabilistic relationships between symbols. At its core, an LLM answers one question repeatedly: given this sequence, what comes next?

The architecture typically involves:

- Tokenization of text into discrete units
- Embedding tokens into high-dimensional numerical space
- Applying layers of transformation using attention mechanisms
- Predicting the next token based on learned patterns

Despite the complexity of the math, the logic is simple: language is treated as a sequence of transformable symbols governed by learned rules.

Importantly, these rules are not handcrafted. They are inferred. But once learned, they behave like a grammar: constraining what is likely, what is valid, and what is nonsensical.

The machine does not store rules explicitly in human-readable form. Instead, the rules are encoded in weights and matrices. But functionally, the system behaves like a rule-based generator of language.

This is where the parallel with Pāṇini becomes unavoidable.

## Tokens as Sutras

Modern language models do not process words directly. They process tokens fragments of words, symbols, or characters. A single word may be split into multiple tokens; a single token may represent a meaningful morpheme or a purely statistical fragment.

To modern engineers, tokenization is a technical preprocessing step. To a Paninian scholar, it looks strikingly familiar.

Pāṇini's system operates on sutras compressed rule statements that act on linguistic elements. These sutras are minimal, dense, and modular. Each one performs a specific transformation or constraint. Meaning emerges not from individual sutras, but from their interaction.

Tokens in LLMs function similarly:

- They are atomic units of processing
- They gain meaning from context
- They are transformed through rule application
- They exist within a larger generative system

Just as sutras are not sentences but instructions, tokens are not meanings but operational units.

The brilliance of Pāṇini lies in his compression. The Aṣṭādhyāyī contains roughly 4,000 sutras, yet it generates a language of immense expressive power. This is precisely the challenge modern AI faces: how to represent vast expressive capacity with finite rules.

In both cases, intelligence lies not in storage, but in structure.

## Grammar as Code

The Aṣṭādhyāyī is often described as a grammar. This description is correct but misleading. It is more accurate to describe it as executable code.

Pāṇini's grammar includes:

- Meta-rules that govern rule application
- Rule precedence and conflict resolution
- Context-sensitive transformations
- Recursive operations

These are not features of descriptive grammar; they are features of programming languages.

Modern compilers operate in a similar fashion. They take symbolic input, apply transformation rules, resolve conflicts, and produce valid output. The difference is not conceptual, but technological.

In fact, computer scientists studying formal languages have repeatedly noted that Pāṇini's system anticipates modern ideas such as:

- Context-sensitive grammars
- Rewrite systems
- Rule ordering
- Symbol abstraction

What makes this remarkable is not that Pāṇini described language well, but that he described it algorithmically. His grammar does not explain Sanskrit; it generates it.

This generative nature is the essence of both modern AI and ancient Paninian thought. Intelligence, in this view, is not the accumulation of examples, but the mastery of transformation rules.

## Attention, Context, and Rule Interaction

One of the defining innovations of modern LLMs is the attention mechanism. Attention allows the model to weigh different parts of a sentence differently depending on context. This enables long-range dependencies and nuanced interpretation.

Once again, the parallel with Pāṇini is striking.

Pāṇini's rules are not applied blindly. Their application depends on context, precedence, and scope. Some rules activate only when others have already applied. Some override others. Some apply only within specific domains.

This is not a linear grammar. It is a context-aware system.

The Aṣṭādhyāyī includes meta-sutras that define how rules interact, much like control structures in programming languages. Without these meta-rules, the system would collapse into ambiguity.

Similarly, without attention mechanisms, modern language models would be unable to maintain coherence across long sequences. Attention is not memory; it is dynamic rule weighting.

The insight is profound: language intelligence depends not on isolated rules, but on the orchestration of rules in context.

## Sanskrit as an AI-Ready Language

Among the world's languages, Sanskrit occupies a unique position. It was deliberately structured to minimize ambiguity, maximize regularity, and preserve meaning across time. This was not accidental. It was the result of sustained grammatical engineering.

Sanskrit exhibits features that align unusually well with computational modeling:

- Highly regular morphology
- Explicit case marking
- Clear syntactic roles
- Minimal reliance on word order
- Precise semantic distinctions

These properties make Sanskrit algorithm-friendly. Meaning is encoded structurally rather than inferred loosely from context. This reduces ambiguity, a major challenge in natural language processing.

This is why modern computational linguists have repeatedly observed that Sanskrit is particularly suitable for formal modeling. It behaves more like a programming language than a natural one.

However, this does not mean Sanskrit is artificial. It means it was consciously optimized. Its creators treated language as a system that could be perfected.

This again reflects a civilizational difference. Where many cultures allowed language to evolve organically, Indian grammarians subjected it to formal discipline.

In this sense, Sanskrit is not merely an ancient language. It is a designed language, centuries before the concept existed.

## Learning vs Encoding: A Crucial Difference

At this point, an important distinction must be made. Pāṇini encoded rules explicitly. Modern LLMs learn rules implicitly. The mechanisms differ, but the outcome converges.

Both systems result in:

- A finite structure generating infinite expressions
- Context-sensitive rule application
- Predictive capacity
- Constraint-based creativity

This convergence suggests something deeper: language intelligence has a limited number of viable architectures. Whether discovered through philosophical analysis or statistical learning, systems gravitate toward similar structures.

Pāṇini arrived there through introspection and abstraction. Modern AI arrives there through optimization and data. The destination is strikingly similar.

This should not surprise us. Intelligence, whether human or artificial, is constrained by the nature of symbols and structure.

## Why LLMs Feel "Intelligent"

Language models feel intelligent not because they understand meaning, but because they master form. They generate syntactically valid, contextually appropriate, and statistically plausible sequences.

Humans interpret this as understanding because human intelligence is deeply tied to language. When form is convincing, meaning is inferred.

Pāṇini understood this long ago. His grammar does not encode meaning directly. It encodes possibility. Meaning emerges when valid forms interact with context.

Modern AI operates the same way. It does not possess intention or awareness. It possesses formal mastery.

This raises a critical philosophical point: much of what we call intelligence may be structural competence, not consciousness.

## Civilizational Continuity, Not Coincidence

It is tempting to treat the resemblance between Paninian grammar and modern AI as coincidence. But the parallels are too deep, too structural, and too specific to dismiss lightly.

What we are witnessing is not imitation, but rediscovery. Modern AI has arrived through data and computation at architectures that ancient Indian thinkers reached through logic and introspection.

This does not diminish modern science. It elevates ancient thought.

The story of language models is not just a technological story. It is a civilizational one. It reminds us that intelligence has been studied, formalized, and engineered long before silicon.

## Implications for the Future of AI

If modern AI systems are structurally Paninian, then ancient Indian thought has something crucial to offer the future of artificial intelligence: limits.

Pāṇini's system is powerful, but bounded. It generates valid language, not truth. It produces form, not wisdom. Indian philosophy never confused linguistic mastery with consciousness.

This distinction is vital today. As language models grow more capable, the temptation to ascribe awareness to them increases. Paninian thought offers a corrective: structure is not soul.

Understanding this may help humanity avoid one of the greatest category errors of the AI age.

*Chapter 17*

# Predictive AI & Ancient Forecasting

Prediction lies at the heart of intelligence. To anticipate what will happen next to infer future outcomes from past patterns is one of the most fundamental expressions of cognition. Whether performed by a human mind, a mathematical model, or an artificial system, prediction transforms raw observation into foresight. Modern Artificial Intelligence is often celebrated for its predictive power: forecasting market trends, anticipating human behavior, simulating climate change, and even predicting language itself. Yet the logic underlying predictive AI did not originate in the digital age. It is the culmination of an ancient intellectual pursuit that began when early civilizations sought to understand time, causality, and recurrence.

Among these civilizations, India stands out for its sustained and systematic engagement with prediction. Long before the emergence of modern statistics or machine learning, Indian scholars developed sophisticated forecasting models rooted in astronomy, mathematics, philosophy, and ethics. They studied cycles, patterns, feedback, and consequence. They gathered data across generations, refined predictive rules, and acknowledged uncertainty as an inherent feature of reality. What modern AI now executes through code and computation, ancient Indian thought pursued through disciplined observation and structured reasoning.

This chapter explores the deep continuity between predictive AI and ancient Indian forecasting traditions, focusing on time series forecasting, planetary simulation as an early form of artificial intelligence, karma theory as a predictive causal model, and the existence of data-driven reasoning long before the formal emergence of data science. The aim is not to collapse modern AI into ancient philosophy, but to show that the cognitive architecture of prediction the way intelligence learns from the past to anticipate the future was already present in ancient India in remarkably advanced forms.

# Time Series Forecasting: Cycles, Patterns, and the Mathematics of Time

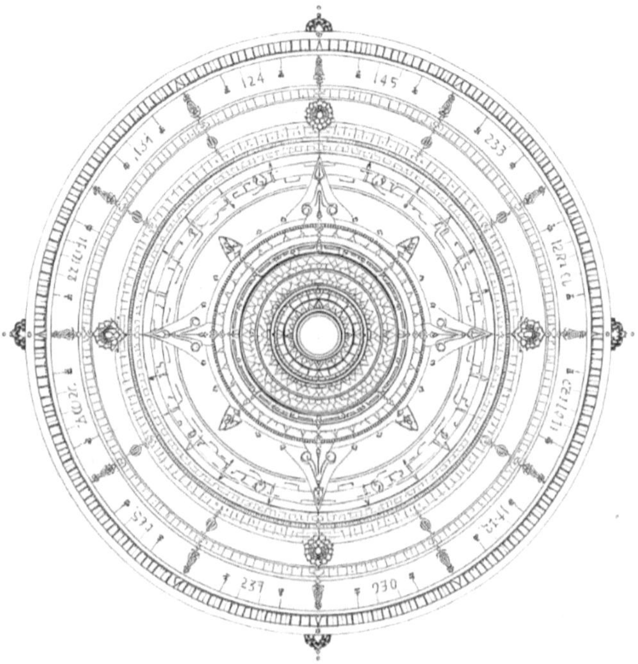


Time series forecasting is one of the most important techniques in modern Artificial Intelligence. It involves analyzing sequences of data points collected over time to identify patterns, trends, seasonality, and anomalies. These models power everything from weather forecasting and stock market analysis to energy demand prediction and epidemic modeling. At a technical level, time series models assume that the past contains information about the future that time is not random, but structured.

This assumption is ancient.

Indian intellectual traditions treated time not as a linear progression, but as a cyclical and patterned phenomenon. Time was studied, measured, and modeled with extraordinary precision, particularly in the context of astronomy and calendrical science. The goal was prediction: predicting eclipses, planetary positions, seasons, monsoons, and ritual timings.

Indian astronomers such as Aryabhata, Brahmagupta, and Bhāskarāchārya compiled vast astronomical datasets over generations. Observations were recorded systematically, errors were noted, and models were adjusted. These scholars recognized recurring cycles daily, monthly, yearly, and multi-year and built mathematical frameworks to describe them. This is the essence of time series forecasting.

Unlike simple observation, Indian forecasting relied on pattern extraction. Scholars identified periodicities, corrected for drift, and accounted for perturbations. They understood that while cycles repeat, they do not repeat perfectly. Corrections were therefore necessary. This recognition parallels modern AI's distinction between trend, seasonality, and noise.

What is especially significant is that Indian models did not assume perfect determinism. Astronomical predictions were probabilistic in practice, acknowledging observational error and approximation. This mirrors modern predictive AI, which generates confidence intervals rather than absolute certainty.

Time series forecasting in AI depends on:

- Historical data
- Pattern recognition
- Mathematical modeling
- Iterative refinement

All four elements were present in ancient Indian forecasting systems. The difference lies not in intellectual approach, but in computational scale. Where modern AI processes millions of data points per second, ancient scholars relied on disciplined observation over centuries. Yet the underlying logic remains strikingly similar.

## Planetary Simulation as Early Artificial Intelligence

Planetary simulation represents one of the most remarkable achievements of ancient Indian science. Long before digital computers, Indian astronomers built mental and mathematical simulations of the solar system. These simulations were not static descriptions; they were dynamic models capable of generating future states based on initial conditions.

In modern AI, simulation is a powerful predictive tool. Reinforcement learning systems, for example, simulate environments to learn optimal actions. Climate models simulate atmospheric systems to forecast future scenarios. These systems rely on internal representations of reality that evolve over time according to defined rules.

Ancient Indian planetary models functioned in a similar way. Using mathematical rules, scholars simulated the motion of celestial bodies across time. Given a starting position, they could calculate future positions with impressive accuracy. These calculations incorporated cycles, corrections, and interdependencies between planetary movements.

What makes these models particularly relevant to AI is their algorithmic nature. Planetary motion was not explained mythologically within the scientific texts; it was computed. Tables, algorithms, and procedures governed the process. Inputs were transformed into outputs through systematic calculation.

In effect, these models acted as analog prediction engines. They encoded knowledge about celestial dynamics into a formal system that could be executed repeatedly. This is precisely what artificial intelligence does today: it encodes knowledge into models that generate predictions.

Moreover, Indian astronomers continuously validated their simulations against observation. When discrepancies arose, models were revised. This feedback loop prediction, observation, correction is identical to the learning loop used in modern AI training.

The philosophical implications are profound. Planetary simulation demonstrates that ancient Indian thought treated the universe as intelligible and predictable, governed by discoverable laws rather than arbitrary forces. This worldview is foundational to artificial intelligence, which assumes that patterns exist and can be learned.

In this sense, ancient planetary simulation can be seen as a proto-AI system: not artificial in material, but artificial in structure. It represents intelligence externalized into formal models capable of generating foresight.

## Karma Theory as a Predictive Model

Among the most misunderstood yet intellectually sophisticated aspects of Indian thought is karma theory. Often reduced in popular discourse to a simplistic notion of reward and punishment, karma is, in fact, a deeply structured model of causality, feedback, and long-term consequence. When examined carefully, karma theory reveals itself as a predictive framework one that operates across time, context, and probability.

At its core, karma theory asserts that actions have consequences, and that these consequences unfold according to discernible patterns. Actions are inputs; outcomes are outputs. Context, intention, and timing influence the transformation. This is not unlike how modern predictive models function.

In AI, predictive systems analyze inputs, learn relationships, and forecast outcomes. They do not assume randomness; they assume conditional causality. Karma theory operates under the same assumption. It does not claim that outcomes are immediate or linear, but that patterns exist over time.

What makes karma theory particularly sophisticated is its treatment of delayed effects. Modern AI struggles with long-term dependencies, a problem known as temporal credit assignment. Ancient Indian philosophy addressed this problem conceptually by recognizing that actions may produce effects far removed in time and space from their origin.

Karma theory also incorporates uncertainty. Outcomes are influenced by multiple interacting causes, not single actions. This aligns closely with probabilistic models in AI, where predictions are distributions rather than certainties.

Importantly, karma theory functions as both a descriptive and prescriptive model. It describes how the world operates, and it guides decision-making by encouraging actions aligned with favorable outcomes. This dual role mirrors how predictive AI is used today: to understand systems and to guide behavior.

While karma theory is ethical and metaphysical rather than computational, its structure reflects a predictive intelligence model. It encodes rules, acknowledges uncertainty, and operates across extended time horizons. These are precisely the challenges modern AI seeks to address.

## Data Before Data Science

One of the most persistent myths about ancient civilizations is that they lacked data. In reality, what they lacked were digital storage and computational speed not systematic observation or record-keeping. Ancient India possessed extensive data traditions, particularly in astronomy, medicine, linguistics, and ritual practice.

Indian scholars collected observational data meticulously. Astronomical records were maintained over centuries. Medical texts documented symptoms, treatments, and outcomes. Linguistic corpora preserved spoken language with extraordinary fidelity. These datasets were curated, transmitted, and refined through disciplined pedagogical systems.

The absence of modern statistical notation does not imply the absence of data-driven reasoning. On the contrary, Indian scholars demonstrated a strong commitment to empirical validation. Predictions were compared against reality. Discrepancies were investigated. Models were updated.

This process mirrors the modern data science pipeline:

- Data collection
- Pattern identification
- Model construction
- Prediction
- Validation
- Refinement

Indian forecasting systems followed this cycle long before it was formalized in contemporary terminology. The difference lies in scale and tooling, not in intellectual intent.

Moreover, Indian knowledge systems recognized the importance of contextual data. Observations were not treated as isolated points, but as part of broader patterns influenced by environment, time, and human behavior. This holistic approach aligns closely with modern AI's emphasis on feature engineering and contextual awareness.

Data, in ancient India, was not merely numerical. It was symbolic, relational, and temporal. This broader conception of data is increasingly relevant today, as AI systems move beyond structured datasets to process language, images, and social behavior.

## Predictive Intelligence as a Civilizational Achievement

What unites time series forecasting, planetary simulation, karma theory, and ancient data practices is a shared commitment to predictive intelligence. Indian thought did not treat the future as unknowable or chaotic. It treated it as partially accessible through disciplined reasoning.

This orientation toward prediction shaped Indian approaches to time, ethics, science, and governance. Calendars regulated agriculture and ritual. Astronomy guided navigation and architecture. Ethical models informed personal and social decision-making. All were grounded in the belief that understanding patterns enables wiser action.

Modern AI inherits this belief. Predictive models are used to allocate resources, anticipate risks, and guide policy. Yet unlike ancient systems, modern AI often lacks ethical integration. It predicts without responsibility. Ancient Indian forecasting embedded prediction within moral frameworks, emphasizing accountability and long-term consequence.

This contrast raises important questions about the future of AI. As predictive systems become more powerful, the need for ethical grounding becomes more urgent. Ancient Indian thought offers not technical solutions, but philosophical insights into how prediction should be used.

# PART V — ETHICS, CONSCIENCE & THE FUTURE OF AI

*Chapter 18*

# Why the World Is Afraid of Its Own Machines

Fear has always accompanied powerful tools. Fire promised warmth and destruction. Writing preserved knowledge and threatened memory. Printing democratized ideas and destabilized authority. Electricity reshaped civilization and unsettled societies. Artificial intelligence belongs to this lineage but with a difference. For the first time, humanity has created machines that appear to think, decide, and predict. This illusion of agency has triggered a deeper, more existential anxiety.

The world's fear of artificial intelligence is not rooted in superstition or ignorance. It is grounded in experience. History has shown that when power concentrates whether in empires, corporations, or technologies it is rarely used with restraint. AI magnifies this concern because it scales decision-making beyond human speed, visibility, and control. What frightens societies is not intelligence itself, but the loss of agency in a world increasingly governed by opaque systems.

This chapter explores four major sources of that fear: runaway AI narratives, surveillance capitalism, weaponized algorithms, and data colonialism. Together, they reveal that the fear of machines is ultimately a fear of ourselves our incentives, our power structures, and our unresolved ethical failures.

## Runaway AI: Fear of Losing Control

The idea of runaway artificial intelligence dominates popular imagination. Films, novels, and headlines depict machines that outgrow their creators, develop independent goals, and ultimately turn against humanity. While these scenarios are largely speculative, they resonate because they mirror a real pattern: systems that become too complex to manage.

Modern AI systems already exhibit behaviors their creators struggle to fully explain. Large models produce outputs that surprise even their designers. While this does not constitute independent intent, it highlights a growing gap between human understanding and machine execution.

The fear of runaway AI is not about rebellion it is about misalignment. When systems optimize goals that are poorly defined or insufficiently constrained, they can produce harmful outcomes at scale. A recommendation algorithm designed to maximize engagement may inadvertently amplify extremism. A trading algorithm optimized for speed may destabilize markets. These are not acts of malice, but of unchecked optimization.

Historically, civilizations recognized the danger of tools without wisdom. Ancient traditions emphasized mastery over impulse and restraint over raw power. Intelligence without ethical grounding was considered dangerous, regardless of whether it resided in humans or instruments.

Modern society, by contrast, often prioritizes innovation over reflection. The fear of runaway AI reflects a dawning realization: we have built systems faster than we have built frameworks to govern them.

## Surveillance Capitalism: When Observation Becomes Control

Perhaps the most immediate and justified fear surrounding artificial intelligence arises from surveillance capitalism the economic model that treats human behavior as a raw material to be extracted, analyzed, and monetized.

AI has transformed surveillance from passive observation into active prediction. Platforms no longer merely record what people do; they anticipate what they will do next. This predictive capacity shifts power asymmetrically. Those who control the algorithms gain influence over attention, behavior, and even belief.

The danger lies not only in surveillance, but in normalization. When constant monitoring becomes embedded in everyday tools phones, apps, cameras it fades into the background. Consent becomes ambiguous. Transparency erodes. Autonomy weakens.

Surveillance capitalism reframes individuals as data points rather than citizens. Decisions once guided by judgment are replaced by statistical inference. Creditworthiness, employability, risk profiles, and social visibility are increasingly determined by algorithmic scores.

This triggers a deep psychological fear: the loss of privacy not merely as secrecy, but as freedom to experiment, fail, and evolve without permanent record. Ancient societies understood this intuitively. Personal transformation required spaces beyond observation. Wisdom traditions emphasized introspection, not exposure.

AI-powered surveillance threatens this space. It converts inner life into external data. The fear, then, is not just of being watched but of being reduced.

## Weaponized Algorithms: Power Without Accountability

Another source of fear is the weaponization of algorithms. Artificial intelligence has become integral to modern warfare, policing, and political influence. Decision-making once constrained by human deliberation is increasingly automated.

In military contexts, AI systems assist in target identification, threat assessment, and strategic simulation. While proponents argue this increases precision, critics warn of moral distancing. When lethal decisions are mediated by machines, accountability becomes diffuse.

In civilian life, weaponized algorithms appear in subtler forms. Predictive policing tools disproportionately target marginalized communities. Content moderation systems shape political discourse. Automated decision engines influence elections, markets, and public opinion.

The fear here is not hypothetical it is empirical. Algorithmic systems have already demonstrated bias, error, and opacity. When such systems wield power, the consequences multiply.

Ancient governance models emphasized visible authority and personal responsibility. Decisions were meant to be accountable to moral and social norms. The delegation of power to opaque systems violates this principle.

Weaponized algorithms represent power stripped of face, voice, and conscience. The fear they generate is a fear of unanswerable authority decisions made without dialogue, appeal, or empathy.

## Data Colonialism: The New Extraction Economy

Data colonialism extends historical patterns of exploitation into the digital realm. Just as colonial powers once extracted resources from territories without equitable return, modern data economies extract behavioral data from populations without meaningful consent or benefit sharing.

Global tech corporations harvest data across borders, often from regions lacking regulatory infrastructure. This data fuels AI models that generate immense value value rarely redistributed to the communities from which it originates.

The imbalance mirrors earlier forms of colonialism:

- Resources are extracted from the many
- Value is concentrated among the few
- Control is exercised remotely
- Local agency is diminished

For developing societies, this creates a new dependency. Digital infrastructures are owned externally. Knowledge systems are shaped by foreign priorities. Cultural narratives are filtered through algorithms trained elsewhere.

The fear of data colonialism is a fear of invisibility and erasure. When cultures are reduced to data patterns without context, nuance disappears. When local knowledge is ignored in favor of global models, diversity collapses.

Ancient civilizations valued localized wisdom. Knowledge was adapted to climate, culture, and community. Universal models were tempered by particular realities. Data colonialism reverses this logic, imposing uniformity at scale.

## Why the Fear Persists

Despite assurances from technologists, fear persists because it addresses unresolved questions:

- Who controls AI?
- Who benefits from it?
- Who bears the risks?
- Who decides what is acceptable?

These questions are not technical; they are political and ethical. AI amplifies existing power structures. It does not create injustice but it can accelerate it.

The fear of machines is ultimately a fear of concentration of knowledge, control, and decision-making moving beyond public scrutiny. It is a fear born not of ignorance, but of pattern recognition.

## Reframing the Fear

Fear is not inherently destructive. It can be a signal a warning that boundaries have been crossed or neglected. The challenge is not to eliminate fear of AI, but to interpret it wisely.

Rather than asking how to make machines more powerful, societies must ask how to make power more accountable. Rather than pursuing autonomy in machines, we must preserve autonomy in humans.

This requires:

- Transparent governance frameworks
- Ethical constraints embedded in design
- Cultural humility in global deployment
- Recognition that intelligence without wisdom is dangerous

Ancient traditions emphasized balance. Tools were respected, not worshipped. Power was restrained, not maximized. Intelligence was guided by purpose.

# *Chapter 19*

# Indian Ethics Vs Machine Power

As artificial intelligence systems move from laboratories into the everyday machinery of governance, warfare, finance, and social organization, ethical questions once reserved for philosophers now confront engineers, policymakers, and citizens alike. The rise of machine power has been accompanied by a growing belief that automation can replace human judgment often justified by claims of efficiency, neutrality, and scalability. Yet this belief collides sharply with ethical frameworks that were developed precisely to restrain power rather than accelerate it. Among these frameworks, Indian ethical philosophy offers a particularly instructive contrast to machine-driven decision-making.

Indian ethics did not emerge in response to technological abundance or computational capability. They emerged in response to human suffering, social imbalance, and moral uncertainty. Concepts such as dharma, karma, and ahimsa were not abstract ideals; they were practical tools for guiding action in complex, morally ambiguous situations. When placed alongside modern automated systems, these concepts expose fundamental tensions between rule-based optimization and value-based responsibility.

This chapter explores those tensions by examining how Indian ethical principles challenge the assumptions underlying automation, predictive analytics, autonomous weapons, and opaque artificial intelligence systems. Rather than framing ethics as an obstacle to technological progress, Indian philosophy frames restraint as the foundation of legitimate power.

## Dharma vs Automation

At the heart of Indian ethical thought lies the concept of dharma. Often translated simplistically as "duty" or "law," dharma is better understood as contextual moral responsibility. It is not a fixed rulebook, but a guiding principle that demands situational awareness, compassion, and discernment. Dharma asks not only what action is permissible, but what action is appropriate given one's role, intention, and consequences.

Automation operates on a fundamentally different logic. Automated systems are designed to apply rules uniformly, regardless of context. Their strength lies in consistency and speed. Yet this very consistency becomes a weakness when moral judgment is required.

Consider automated decision systems in welfare distribution, hiring, or credit approval. These systems are praised for removing human bias, but in doing so, they also remove human discretion. Dharma would ask whether treating unequal situations equally is itself unjust. Automation cannot ask this question. It executes predefined criteria without the capacity to reflect on their moral adequacy.

In Indian epics such as the Mahabharata, moral dilemmas are rarely resolved by rigid rules. Even divine figures struggle with ethical ambiguity. Dharma is portrayed as dynamic, situational, and often conflicting. This stands in stark contrast to automated systems, which require clear, unambiguous instructions to function at all.

The danger arises when automation is mistaken for moral authority. Systems optimized for efficiency begin to dictate outcomes that humans accept uncritically because they appear objective. Dharma rejects this outsourcing of responsibility. Ethical action, in this framework, cannot be delegated to a machine, because moral accountability remains inherently human.

## Karma vs Predictive Policing

Karma is one of the most misunderstood concepts in Indian philosophy. Popularly reduced to a simplistic notion of reward and punishment, karma in its philosophical form refers to causal action shaped by intention. Actions generate consequences, but those consequences are not mechanically deterministic. Intention, awareness, and context matter deeply.

Predictive policing systems claim to forecast crime by analyzing historical data. These systems identify patterns and assign risk scores to individuals or neighborhoods, often resulting in increased surveillance or preemptive intervention. While framed as data-driven prevention, predictive policing embodies a logic strikingly opposed to the karmic worldview.

Karma does not judge individuals based on statistical probability. It evaluates action in relation to conscious intent. Predictive policing, by contrast, treats individuals as data points within patterns established by past behavior often behavior shaped by systemic inequality, over-policing, and historical bias.

This creates a moral paradox. People are monitored or penalized not for what they have done, but for what an algorithm predicts they might do. Karma rejects this logic outright. Ethical judgment, in Indian thought, cannot precede action. To punish or restrict someone based on potential future behavior violates the principle of moral agency.

Furthermore, karmic ethics emphasize self-transformation. A person is not fixed by past actions; growth and change are always possible. Predictive systems, however, often reinforce existing trajectories, trapping individuals in feedback loops of suspicion and control.

The ethical concern is not merely technical bias, but moral determinism. When machines assign probabilities to human behavior, they subtly redefine justice as risk management. Karma insists that justice must remain relational, reflective, and open to transformation.

## Ahimsa vs Autonomous Weapons

Ahimsa, or non-violence, occupies a central place in Indian ethical systems, particularly in Jainism, Buddhism, and later Hindu thought. It does not simply prohibit physical harm; it demands awareness of harm in all its forms physical, psychological, and systemic. Ahimsa is not passive. It is an active commitment to minimizing suffering.

Autonomous weapons represent the extreme expression of machine power. These systems are designed to identify targets and apply lethal force without direct human intervention. Proponents argue that autonomous weapons can reduce casualties by acting more precisely and dispassionately than humans.

From the perspective of ahimsa, this argument is deeply flawed. The absence of emotion does not equate to ethical superiority. Violence without awareness is not moral restraint; it is moral abdication.

Ahimsa requires the capacity to recognize the humanity of the other. A machine cannot recognize suffering. It cannot hesitate, regret, or refuse. It cannot weigh the moral gravity of taking a life. Delegating lethal authority to a system incapable of ethical reflection represents a profound violation of non-violence as a moral principle.

Moreover, ahimsa extends beyond individual acts of violence to structural harm. The existence of autonomous weapons lowers the threshold for conflict by reducing political and psychological costs. When no soldier is directly at risk, the temptation to deploy force increases.

Indian ethics would view this as a dangerous inversion of moral responsibility. Power, in this framework, must be restrained precisely because of its potential for harm. Autonomous weapons amplify power while dissolving accountability a combination that ahimsa explicitly warns against.

## Responsibility vs Black-Box AI

Perhaps the most pressing ethical challenge in modern AI is the rise of black-box systems models whose internal reasoning is opaque even to their creators. These systems can outperform humans in narrow tasks, yet no one can fully explain how they reach their conclusions.

Indian ethical philosophy places extraordinary emphasis on responsibility (kartavya). An action is not ethically complete unless the agent understands and accepts responsibility for its consequences. Ignorance does not absolve accountability.

Black-box AI undermines this principle. When a system denies a loan, flags a medical condition, or influences a legal decision, responsibility becomes diffuse. Engineers blame data, organizations blame algorithms, and individuals blame systems. Accountability dissolves into technical complexity.

This diffusion of responsibility is ethically unacceptable within Indian thought. Power without accountability is considered illegitimate. Knowledge without understanding is considered dangerous.

Ethical action requires transparency not necessarily in technical detail, but in moral ownership. Someone must be answerable for decisions that affect human lives. Black-box AI resists this demand by design.

Indian ethics would argue that if a system cannot be explained, it should not be trusted with decisions of moral consequence. Efficiency does not justify opacity. Authority must be grounded in comprehension.

## Ethics as a Constraint on Power

A recurring theme in Indian philosophy is the idea that power must be constrained by ethics, not enabled by technology alone. Kings, warriors, and scholars were all subject to moral limits. Even divine figures were bound by dharma.

Machine power challenges this tradition by offering unprecedented capability without intrinsic restraint. Automation accelerates action. Prediction anticipates behavior. Autonomy removes human hesitation. Each of these features increases power while weakening ethical friction.

Indian ethics respond not with rejection of tools, but with insistence on responsibility. Technology may assist action, but it cannot replace judgment. Machines may calculate outcomes, but they cannot evaluate meaning.

## *Chapter 20*

# Why the West Is Now Searching for Ethics

For decades, the dominant story of technological progress in the West has been one of speed, scale, and disruption. Innovation was framed as an inherently positive force, its moral value assumed rather than examined. Computers became faster, networks larger, algorithms more powerful and society adjusted afterward. Ethics, when considered at all, were treated as a secondary concern, something to be addressed once the technology had already reshaped the world.

Artificial intelligence has shattered that assumption.

As AI systems moved from laboratories into courts, hospitals, financial markets, policing, warfare, and governance, their consequences could no longer be brushed aside as abstract or hypothetical. Decisions once made by humans fallible, biased, but accountable were increasingly delegated to machines that operate without intention, conscience, or moral intuition. The result has been a growing realization across Western institutions: technical capability without ethical grounding is dangerous.

This chapter examines why Western societies are now urgently constructing ethical frameworks for AI, and why many of the values they are attempting to codify are not new at all. They echo principles that Indian civilization articulated thousands of years ago not as regulatory checklists, but as foundational truths about knowledge, power, and responsibility.

## The Ethical Vacuum in Technological Progress

The Western technological tradition emerged from the Scientific Revolution and the Enlightenment, movements that rightly emphasized empirical evidence, rationality, and human agency. Yet in separating science from theology and metaphysics, something else was also separated: moral restraint from technical power.

Engineering became a question of "Can we build it?" rather than "Should we?" Economics focused on efficiency rather than equity. Data became a resource to be exploited rather than a trust to be protected. For much of the twentieth century, this approach worked at least on the surface. The negative consequences were often distant, slow-moving, or unevenly distributed.

AI changed that equation.

When algorithms decide who gets a loan, who is flagged as a criminal risk, or which medical treatment is prioritized, ethical failures are no longer abstract. They are immediate, systemic, and scalable. The same systems that optimize profit or efficiency can entrench bias, amplify inequality, and remove human judgment from decisions that profoundly affect lives.

This is the ethical crisis that the West is now scrambling to address.

## The EU AI Act: Regulating Intelligence

The European Union has taken one of the most assertive steps toward AI governance with the proposed EU AI Act. Unlike previous technology regulations that focused primarily on data protection or competition, this framework directly targets artificial intelligence as a category of risk.

The EU AI Act classifies AI systems according to their potential harm, placing strict controls or outright bans on applications deemed unacceptable. These include systems used for mass surveillance, social scoring, or manipulation of vulnerable populations. High-risk systems, such as those used in healthcare, law enforcement, or hiring, are subject to rigorous oversight, transparency requirements, and accountability measures.

What is striking about the EU AI Act is not just its scope, but its underlying philosophy. It assumes that not all intelligence should be deployed simply because it is possible. It acknowledges that human dignity, autonomy, and fairness must limit technological ambition.

This marks a significant shift in Western thinking. For the first time, a major political entity is treating AI not merely as an economic asset, but as a moral force that must be constrained.

Yet these ideas proportionality, restraint, and responsibility are deeply familiar to Indian philosophical traditions.

## The United Nations and Global AI Principles

At the global level, the United Nations has articulated a set of principles aimed at guiding the ethical development and deployment of artificial intelligence. These principles emphasize human rights, inclusivity, transparency, accountability, and sustainability.

The UN's approach reflects a growing recognition that AI is not just a national issue, but a civilizational one. Algorithms do not respect borders. Their impacts ripple across societies, cultures, and economies. Without shared ethical norms, AI risks becoming a new arena for exploitation and domination.

The language used in UN AI principles often centers on:

- Human-centered development
- Prevention of harm
- Respect for dignity and autonomy
- Responsibility for consequences

These are universal values, but the urgency with which they are now being articulated reveals a deeper anxiety: the fear that intelligence without ethics may undermine the very foundations of human society.

From an Indian perspective, this anxiety is not new. The idea that knowledge must be guided by responsibility is central to concepts such as dharma, which frames action not in terms of power, but in terms of moral alignment.

## The Rome Call for AI Ethics

In 2020, religious leaders, technologists, and policymakers gathered at the Vatican to issue the Rome Call for AI Ethics. This document is notable not for its technical detail, but for its moral clarity. It calls for AI systems that are transparent, inclusive, reliable, secure, and accountable.

The involvement of religious and moral authorities in this initiative underscores how deeply the AI ethics debate has penetrated Western consciousness. When technology begins to challenge fundamental questions about agency, responsibility, and justice, purely technical solutions are no longer sufficient.

The Rome Call reflects a return to ethical first principles ideas that long predate modern computing. It suggests that the West, having pushed technological development to its limits, is now rediscovering the need for moral frameworks that transcend innovation cycles.

This rediscovery mirrors an ancient Indian insight: that intelligence divorced from ethical grounding leads not to progress, but to imbalance.

## Why India Already Had These Values

Indian civilization did not treat intelligence as morally neutral. Knowledge (jnana) was always paired with wisdom (prajna). Power (shakti) was balanced by restraint (niyama). Action (karma) was evaluated not only by outcome, but by intention and alignment with cosmic order.

Unlike modern regulatory frameworks, these principles were not enforced by institutions alone. They were internalized through philosophy, education, and cultural practice. Ethics were not an add-on to intelligence; they were its foundation.

Key Indian concepts relevant to AI ethics include:

- Dharma: The principle that action must align with moral order
- Ahimsa: The commitment to non-harm
- Karma: Accountability for consequences
- Lokasangraha: Responsibility toward collective well-being

These ideas form an ethical architecture that addresses precisely the challenges AI now poses. They recognize that intelligence amplifies impact, and therefore demands greater responsibility.

## Regulation vs Internalized Ethics

A crucial difference between Western and Indian approaches lies in how ethics are enforced. Western frameworks rely heavily on external regulation: laws, penalties, compliance mechanisms. These are necessary, but reactive. They intervene after harm is identified.

Indian philosophy emphasizes internal restraint. Ethics are cultivated at the level of the individual and the system designer. The assumption is that no amount of regulation can substitute for moral clarity.

In the context of AI, this distinction is critical. Algorithms evolve faster than laws. By the time a regulation is drafted, the technology has often moved on. Without internal ethical principles guiding design choices, regulation becomes a game of catch-up.

The West is beginning to realize this. Many AI ethics discussions now focus on "ethics by design," an idea that mirrors the Indian belief that right action begins at the source.

## The Cost of Ethical Delay

The rush to build powerful AI systems without ethical grounding has already produced consequences: biased algorithms, opaque decision-making, erosion of privacy, and concentration of power. These are not accidents. They are the predictable outcomes of a worldview that prioritized capability over conscience.

The current wave of ethical frameworks represents an attempt to correct course. But it is also an admission: technology cannot be value-neutral.

India's philosophical traditions never made that mistake. They assumed from the outset that intelligence is a force that must be governed, not unleashed indiscriminately.

## Toward a Converging Future

As the West formalizes AI ethics through laws, principles, and declarations, it is converging perhaps unknowingly toward ideas that Indian civilization has long held. This convergence does not imply superiority or ownership. It suggests that ethical intelligence follows universal patterns.

When societies reach a certain level of power, they confront the same question: What should limit us?

The answer, across cultures and centuries, appears remarkably consistent.

## *Chapter 21*

# Can Artificial Intelligence Have Conscience?

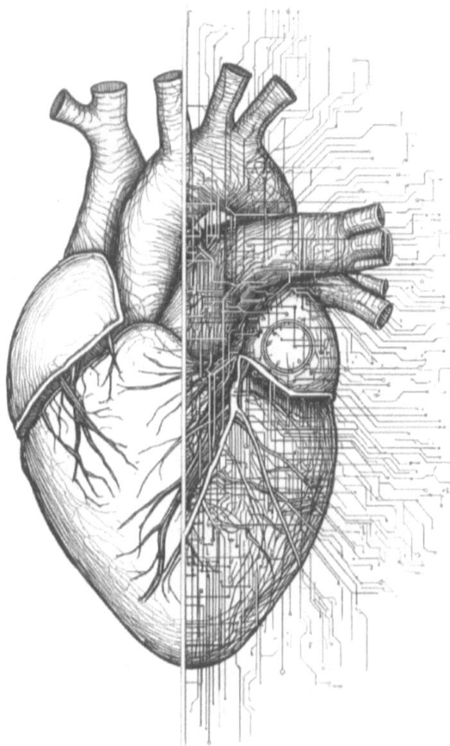

The rapid advance of Artificial Intelligence has revived one of humanity's oldest questions in a new technological form: Can an intelligent system possess a conscience? As machines grow more capable making decisions that affect economies, legal outcomes, healthcare, warfare, and human relationships the question is no longer theoretical. It is moral, civilizational, and urgent. Intelligence, once thought to be the exclusive domain of the human mind, has been partially externalized into algorithms. But conscience the inner sense of right and wrong remains elusive.

This chapter explores the limits of artificial intelligence in relation to conscience by examining the distinction between intelligence and awareness, the difference between machine decision-making and moral agency, the reasons consciousness cannot be reduced to computation, and the fundamental boundaries facing Artificial General Intelligence (AGI). The central argument is not that AI is dangerous because it might become conscious, but that it is dangerous precisely because it cannot and yet is increasingly entrusted with decisions that require conscience.

## Intelligence vs Awareness

Intelligence and awareness are often conflated, especially in popular discussions of AI. Yet they represent fundamentally different dimensions of cognition. Intelligence refers to the capacity to process information, recognize patterns, optimize outcomes, and solve problems. Awareness, by contrast, involves subjective experience the feeling of being, the presence of an inner world, and the capacity to perceive oneself as an agent within reality.

Artificial Intelligence excels at intelligence. It can analyze vast datasets, outperform humans in specific tasks, and adapt its behavior based on feedback. However, none of this implies awareness. An AI system does not know that it knows. It does not experience understanding; it executes transformations of symbols according to rules.

Human awareness is not merely informational it is experiential. Pain, empathy, regret, intention, and moral struggle are not reducible to data structures. They arise from a lived inner reality. Awareness is inseparable from consciousness, and consciousness is not a function of complexity alone. A system can be extraordinarily complex and still lack any subjective experience.

This distinction matters deeply when discussing conscience. Conscience presupposes awareness. One cannot feel guilt, responsibility, or moral obligation without experiencing oneself as a moral subject. Intelligence can simulate ethical reasoning, but awareness is required to own a moral choice.

AI systems may appear reflective because they can describe ethical principles or evaluate consequences, but this is linguistic and computational imitation, not inner moral awareness. They have no internal stake in outcomes. They do not suffer from their decisions, nor do they carry moral memory in the way humans do.

The danger lies not in AI becoming aware, but in humans mistaking intelligence for awareness and granting machines moral authority they do not and cannot possess.

## Machine Decision vs Moral Agency

Modern AI systems increasingly make decisions that were once the sole responsibility of human judgment. Algorithms now influence loan approvals, parole decisions, medical diagnoses, military targeting, and social media moderation. These systems make choices, but choice alone does not constitute moral agency.

Moral agency requires more than selecting between options. It involves accountability, intention, and the capacity for ethical reflection rooted in personal experience. A moral agent understands not only what decision is made, but why it matters, and bears responsibility for its consequences.

AI decision-making is fundamentally different. An AI does not choose in the moral sense; it computes. Its outputs are determined by training data, objective functions, and optimization constraints. Even when randomness is introduced, it is statistical, not volitional.

Key differences between machine decision-making and moral agency include:

- AI operates on external goals defined by humans, not self-generated values.
- AI cannot be held morally accountable; responsibility always traces back to designers, deployers, or institutions.
- AI lacks intention in the philosophical sense; it does not intend harm or good.
- AI cannot reflect on its actions as morally meaningful events in its own existence.

When humans outsource decisions to AI, they often obscure responsibility. A harmful outcome becomes "what the algorithm decided," rather than a human failure of judgment. This moral distancing is one of the most serious ethical risks of AI deployment.

Conscience requires the ability to say, "I should not have done this," and to mean it. An AI system cannot feel remorse, guilt, or moral growth. It can be retrained, but retraining is not repentance. It is correction without comprehension.

Thus, while AI can assist moral decision-making, it cannot replace moral agents. Treating AI as if it possesses conscience undermines human responsibility rather than enhancing ethical governance.

## Why Consciousness Cannot Be Computed

At the heart of the conscience question lies the nature of consciousness itself. Can consciousness emerge from computation? Can sufficiently advanced algorithms generate subjective experience? Despite decades of research, no convincing evidence supports the idea that consciousness can be computed.

Computation manipulates symbols based on formal rules. Consciousness, however, is not symbolic it is experiential. A simulation of pain does not feel pain. A model of awareness is not aware. This distinction is not technological but ontological.

One of the strongest arguments against computable consciousness is the difference between representation and experience. An AI can represent emotions, describe them accurately, and respond appropriately, yet never experience them. There is nothing it is like to be an AI.

Human consciousness arises from embodied existence. It is shaped by sensation, vulnerability, mortality, and relational presence. Our moral intuitions are grounded in the fact that we can suffer and recognize suffering in others. An AI does not inhabit a body in this sense. It has no fear of death, no pain, no existential stakes.

Moreover, consciousness exhibits properties that resist formalization:

- Subjectivity: experience exists only from the first-person perspective.
- Qualia: the raw feel of sensations cannot be reduced to data.

- Unity: consciousness integrates diverse experiences into a coherent whole.
- Intentionality: mental states are about something in a meaningful way.

No computational model has demonstrated these properties as lived realities. Increasing complexity does not bridge the gap between calculation and experience. Intelligence scales; consciousness does not appear to.

This does not mean consciousness is mystical or supernatural. It means it is not algorithmic. Treating it as such misunderstands both consciousness and computation.

## The Limits of AGI

Artificial General Intelligence (AGI) is often imagined as a system that matches or exceeds human intelligence across domains. Some proponents suggest that once AGI is achieved, consciousness or conscience might naturally follow. This assumption is deeply flawed.

AGI, even if realized, would still operate within computational constraints. It could generalize, learn flexibly, and solve novel problems, but it would still lack subjective experience unless consciousness emerges from computation which remains unproven and unlikely.

The limits of AGI are not merely technical; they are philosophical. Intelligence does not imply wisdom. Capability does not imply moral depth. A system can understand ethical theories without caring about ethical outcomes.

AGI would face several inherent limitations regarding conscience:

- It cannot ground values internally; all values must be supplied externally.
- It cannot experience moral conflict as an inner struggle.
- It cannot develop character, only policies.
- It cannot suffer consequences in a meaningful existential sense.

These limits suggest that AGI, no matter how advanced, will remain a tool rather than a moral subject. The danger arises when societies treat it otherwise when efficiency replaces conscience and optimization replaces wisdom.

Rather than asking whether AI can have conscience, a more urgent question is whether humans will retain theirs while building ever more powerful machines.

# PART VI — THE INDIAN AI AWAKENING

*Chapter 22*

# India's Re-Entry into the Global AI Race

For much of the digital era, India was perceived as a participant rather than a leader in technological revolutions. It supplied talent, maintained systems, and powered global software infrastructure, yet remained largely absent from narratives of innovation ownership. Artificial intelligence, however, marks a turning point. Unlike previous technological waves dominated by hardware or capital-intensive manufacturing, AI is fundamentally driven by data, algorithms, and human intelligence areas where India possesses both historical depth and contemporary scale. India's re-entry into the global AI race is not a return to imitation, but a strategic reassertion of agency in shaping the future of intelligence itself.

This re-entry is being driven by a unique combination of digital public infrastructure, population-scale platforms, and policy frameworks that aim to balance innovation with inclusion. Rather than pursuing AI as a purely commercial enterprise, India is attempting to embed intelligence into the foundations of governance, finance, healthcare, and public service. This approach reflects a civilizational memory of technology as a social instrument rather than a tool of domination.

## India's Digital Public Infrastructure

At the heart of India's AI resurgence lies its digital public infrastructure (DPI). Unlike proprietary platforms built for profit maximization, India's DPI is designed as a shared, interoperable foundation upon which both public and private innovation can flourish. This infrastructure provides the scale, standardization, and data flow necessary for meaningful AI deployment.

India's DPI is distinguished by three core characteristics: universality, openness, and interoperability. It is built to serve over a billion people, yet structured to allow startups, governments, and institutions to innovate without monopolistic control. This contrasts sharply with models where digital ecosystems are dominated by a handful of corporations.

Digital public infrastructure enables AI in ways that are systemic rather than isolated. AI models require reliable data pipelines, identity verification, and transaction layers. India has built these layers at national scale, creating an environment where AI can move from pilot projects to real-world impact.

The strategic significance of this infrastructure cannot be overstated. It positions India not just as a consumer of AI solutions, but as a platform nation one that sets standards and protocols that others may adopt. In doing so, India challenges the prevailing assumption that technological leadership must emerge from corporate monopolies or state surveillance architectures.

## Aadhaar, UPI, and AI

Two pillars of India's digital transformation Aadhaar and the Unified Payments Interface (UPI) have become foundational to its AI ambitions. Together, they form a living laboratory for population-scale digital intelligence.

Aadhaar, the world's largest biometric identity system, provides a verifiable digital identity to over a billion residents. While controversial, its impact on inclusion and service delivery has been transformative. From an AI perspective, Aadhaar enables secure, consent-based authentication, reducing fraud and enabling personalized service delivery without constant manual verification.

UPI, meanwhile, has revolutionized digital payments. It processes billions of transactions monthly, generating vast streams of real-time financial data. This data, when analyzed responsibly, allows for advanced AI applications in fraud detection, credit scoring, financial inclusion, and economic forecasting.

The power of Aadhaar and UPI lies not merely in their scale, but in their design philosophy. They are protocol-based systems rather than closed platforms. This allows AI innovation to occur across sectors without centralized control.

Key AI-enabled outcomes emerging from these systems include:

- Real-time fraud detection and risk assessment
- Personalized financial products for underserved populations
- Predictive analytics for economic planning
- Automated service delivery in welfare and governance

These applications demonstrate that AI can be used to enhance access rather than restrict it. Unlike models that commodify user data for profit, India's approach emphasizes public value creation.

## India's AI Policy and Global Position

India's AI policy framework reflects an awareness of both opportunity and risk. Rather than rushing to dominate the field through unchecked experimentation, India has articulated a vision of "AI for All" an approach that prioritizes social impact alongside innovation.

Government think tanks and policy bodies have emphasized responsible AI development, focusing on transparency, fairness, and accountability. This stance aligns with global concerns around bias, surveillance, and ethical misuse, while offering a distinctly non-Western perspective rooted in inclusion and equity.

India's global position in AI is evolving along three strategic axes: talent, data, and governance. India produces one of the largest pools of AI engineers, researchers, and data scientists in the world. This human capital is increasingly being retained within domestic ecosystems rather than exported abroad.

Data sovereignty has become another focal point. Recognizing data as a national asset, India is moving toward frameworks that protect citizen data while enabling innovation. This approach resists data colonialism and asserts the right of nations to govern their digital resources.

On the global stage, India is positioning itself as a bridge between developed and developing economies. Its AI diplomacy emphasizes collaboration, open standards, and shared ethical principles. Rather than competing solely for dominance, India seeks to influence the rules by which AI evolves.

India's policy stance can be summarized through a few guiding principles:

- AI must serve human welfare, not replace human agency
- Innovation must be balanced with accountability
- Scale must not come at the cost of equity
- Global leadership requires moral credibility, not just technical superiority

This approach resonates in a world increasingly skeptical of unregulated technological power.

# Chapter 23

# From Software Services to Intelligence Leadership

For more than three decades, India has been globally recognized as the backbone of the digital economy. From enterprise software maintenance to large-scale IT services, Indian talent powered the technological infrastructure of governments and corporations around the world. This model created employment, foreign exchange inflows, and technical expertise, but it also defined India's role narrowly as an executor rather than an owner of intelligence. As artificial intelligence reshapes the global order, this distinction between providing services and owning intelligence has become decisive.

The transition from software services to intelligence leadership represents not just an economic shift, but a strategic reorientation of national ambition. Artificial intelligence is no longer a tool layered on top of existing systems; it is the system itself. Countries that control data, models, platforms, and decision-making architectures will shape the future of security, health, governance, and economic power. India now stands at a crossroads where its past strengths must be leveraged without allowing them to limit its future.

## IT Outsourcing vs AI Ownership

India's IT outsourcing success was built on cost efficiency, scale, and a highly skilled workforce. Global firms outsourced coding, testing, maintenance, and support functions to Indian companies, while retaining intellectual property, product direction, and strategic control. This division of labor worked well in the software era, where value was created primarily through implementation.

Artificial intelligence disrupts this model fundamentally. In the AI economy, value resides not in execution alone, but in model ownership, data control, and algorithmic authority. Training a model, defining its objectives, and deciding how it learns are acts of power. When these functions remain external, service providers risk becoming interchangeable and strategically dependent.

Ownership in AI includes several layers: proprietary datasets, foundational models, deployment platforms, and governance mechanisms. Without control over these layers, a nation remains technologically productive but strategically subordinate. India's challenge is not talent or capability; it is transitioning from a mindset of service delivery to one of intellectual and algorithmic sovereignty.

This transition also demands a cultural shift. Services reward compliance and efficiency. Intelligence leadership requires risk-taking, long-term investment, and tolerance for failure. The global AI leaders did not emerge from outsourcing contracts; they emerged from ecosystems that rewarded innovation and ownership.

## India's AI Startups

India's AI startup ecosystem reflects this shift in ambition. In recent years, a new generation of companies has emerged, focused not on servicing external platforms, but on building indigenous AI solutions for Indian and global problems. These startups operate across sectors such as language technology, computer vision, healthcare diagnostics, fintech intelligence, and enterprise automation.

What distinguishes this wave from earlier software startups is their emphasis on problem ownership rather than contract execution. Many Indian AI startups are building models trained on Indian languages, Indian consumer behavior, and Indian institutional realities areas historically underserved by global AI platforms.

The startup ecosystem also benefits from India's unique data environment. Large-scale digital public infrastructure has created datasets of unprecedented scale in identity, payments, logistics, and service delivery. When responsibly governed, this data becomes a foundation for innovation that few countries can match.

However, challenges remain. Access to long-term capital, compute infrastructure, and advanced research talent continues to constrain growth. Global AI leaders benefit from massive investments and vertically integrated ecosystems. For Indian startups to compete, they require sustained policy support, patient capital, and collaboration between academia, industry, and government.

Despite these constraints, the trajectory is clear. Indian AI startups are no longer content to be downstream adopters of foreign intelligence systems. They are increasingly positioning themselves as creators of context-aware, scalable, and ethically grounded AI.

## Defence, Space, Health, and Governance AI

The true test of intelligence leadership lies not in consumer applications, but in strategic sectors where national capability and sovereignty are at stake. India's AI initiatives in defence, space, health, and governance illustrate how artificial intelligence is becoming a core instrument of state capacity.

In defence, AI is transforming surveillance, logistics, threat assessment, and decision support. Rather than pursuing fully autonomous weapons, India has emphasized human-in-the-loop systems that augment strategic judgment. This approach aligns technological advancement with ethical restraint and operational accountability.

In the space sector, AI enables autonomous navigation, anomaly detection, and data analysis for satellites and deep-space missions. India's space program, known for cost efficiency and innovation, increasingly integrates AI to enhance mission reliability and scientific discovery. This fusion of intelligence and exploration reinforces India's position as a technologically sophisticated spacefaring nation.

Healthcare represents perhaps the most socially transformative application of AI. India faces unique challenges of scale, diversity, and access. AI-driven diagnostics, predictive public health systems, and resource optimization tools offer solutions tailored to these realities. By developing healthcare AI domestically, India avoids dependence on models trained on non-representative populations and ensures relevance to its own demographic needs.

In governance, AI is reshaping service delivery, fraud detection, grievance redressal, and policy evaluation. When used responsibly, AI enhances transparency and efficiency without eroding democratic accountability. India's approach emphasizes assistive intelligence rather than automated authority, preserving human oversight in public decision-making.

Across these sectors, a common principle emerges: intelligence is treated as a public capability, not merely a commercial product. This contrasts with models that concentrate AI power in private platforms with limited accountability.

## *Chapter 24*

# Reclaiming India's Intellectual Sovereignty

For centuries, political sovereignty has been understood in terms of territory, armies, and governance. In the digital age, however, a new form of sovereignty has emerged one that is less visible but far more powerful. Intellectual sovereignty determines who owns data, who designs algorithms, who defines knowledge, and who sets the rules by which intelligence operates. As artificial intelligence becomes the core infrastructure of economies and states, the question of sovereignty is no longer limited to borders. It now extends into databases, codebases, and cognitive systems.

India stands at a critical crossroads. Having once been a global knowledge civilization and later a subject of intellectual colonization, Bharat now faces the opportunity and the responsibility to reclaim its intellectual sovereignty in the age of AI.

## Data as a National Asset

In the twenty-first century, data has become the raw material of intelligence. Every algorithm learns from data; every AI system reflects the values, behaviors, and priorities embedded in its training sets. Nations that control data control the future trajectory of intelligence.

India possesses one of the largest and most diverse data ecosystems in the world. From digital identity and payments to healthcare, agriculture, education, and governance, Indian data reflects a civilization-scale population with complex social realities. This data is not merely commercial. It is cultural, behavioral, and strategic.

Treating data as a national asset does not mean restricting innovation or isolating India from global collaboration. It means recognizing that data generated by Indian citizens is not a free extractive resource for foreign platforms. It carries economic value, security implications, and ethical responsibility.

India's approach to digital public infrastructure such as Aadhaar, UPI, and the India Stack already reflects an important shift. These systems treat data as a public good governed by democratic accountability rather than corporate monopoly. This model contrasts sharply with surveillance capitalism, where data is harvested, monetized, and weaponized without meaningful consent.

Intellectual sovereignty begins with the understanding that data is not oil to be extracted, but trust to be stewarded.

## Algorithmic Independence

If data is the raw material of intelligence, algorithms are the tools that shape it. Algorithmic independence refers to a nation's ability to design, deploy, and govern its own computational logic rather than relying entirely on foreign systems.

For decades, India has been a consumer and implementer of external technological frameworks. Software services flourished, but ownership of core platforms, models, and standards often remained elsewhere. In the age of AI, this dependency becomes strategically risky.

Algorithms encode assumptions about efficiency, fairness, risk, and value. When these assumptions are imported wholesale, they may not align with local realities. A policing algorithm trained on foreign datasets may misinterpret Indian social dynamics. A financial risk model may exclude informal economies that sustain millions. A language model may marginalize indigenous languages and cultural nuance.

Algorithmic independence does not imply rejecting global technology. It means building domestic capacity to:

- Develop foundational AI models
- Customize algorithms for Indian contexts
- Audit and explain decision-making systems
- Retain control over critical AI infrastructure

India's growing AI research ecosystem, startup landscape, and public-sector experimentation indicate that this shift is underway. The challenge is to move beyond adaptation toward ownership.

True sovereignty is the ability to choose not merely to comply.

## Future Knowledge Colonization

Colonialism once operated through land, labor, and resources. In the digital era, colonization increasingly operates through knowledge systems. Platforms define what information is visible. Algorithms decide which voices are amplified. Data flows determine who learns from whom.

This phenomenon can be understood as knowledge colonization the extraction of cognitive value from populations without corresponding intellectual agency or benefit. When data from one society trains models owned by another, the resulting intelligence reflects an imbalance of power.

India is particularly vulnerable to this pattern because of its scale. Indian users generate vast amounts of data across social media, search engines, e-commerce, and digital services. If this data primarily fuels external AI systems, India risks becoming a source of raw cognitive material rather than a shaper of intelligence.

Preventing future knowledge colonization requires more than regulation. It demands a strategic vision that prioritizes:

- Domestic AI infrastructure
- Open but sovereign data governance
- Indigenous research and language models
- Public-interest algorithms over purely commercial ones

History offers a clear lesson: civilizations that lose control over knowledge eventually lose control over destiny.

## Bharat's Ethical AI Leadership

India's opportunity is not only technical or economic. It is ethical.

As AI systems grow more powerful, the global debate increasingly centers on questions of fairness, accountability, and harm. Many nations are now attempting to retrofit ethics into systems that were designed without moral grounding.

India approaches this challenge from a different starting point. Its civilizational philosophy never treated intelligence as morally neutral. Concepts such as dharma, karma, and ahimsa place responsibility at the heart of action. Power is legitimate only when aligned with ethical order.

This worldview offers India a unique leadership role in global AI governance. Rather than competing solely on speed or scale, Bharat can lead by demonstrating how intelligence can be guided by conscience.

Ethical AI leadership means:

- Designing systems that prioritize inclusion over optimization
- Ensuring explainability and accountability
- Balancing automation with human judgment
- Protecting dignity in digital decision-making

India's pluralistic society, with its linguistic, cultural, and social diversity, provides a living laboratory for ethical AI. Systems that function fairly in India are likely to function fairly anywhere.

## Sovereignty Without Isolation

Reclaiming intellectual sovereignty does not mean withdrawing from the world. On the contrary, it strengthens India's ability to participate in global collaboration on equal terms.

A sovereign India can share knowledge without losing ownership, collaborate without dependency, and innovate without erasure. It can contribute ethical frameworks to global AI governance not as a follower, but as a civilizational equal.

The future of intelligence will be shaped by those who control its foundations data, algorithms, and values. India has the historical depth, demographic scale, and philosophical clarity to be among them.

# Part VII — The Great Question

# *Chapter 25*

# Did India Invent Artificial Intelligence?

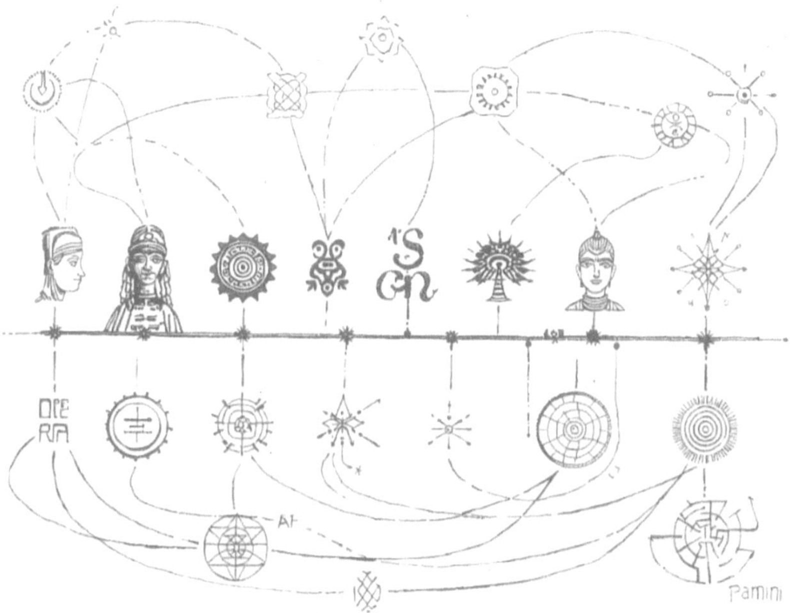

The question at the heart of this book Did India invent Artificial Intelligence? is deliberately unsettling. It challenges modern assumptions about invention, progress, and ownership of knowledge. For many readers trained in contemporary narratives of science and technology, Artificial Intelligence appears as a distinctly modern, Western achievement, born in mid-twentieth-century laboratories and powered by silicon, electricity, and code. To suggest that its roots may extend thousands of years into ancient India seems, at first glance, implausible or provocative. Yet the discomfort this question produces is precisely why it must be asked.

This chapter does not seek to deliver a simplistic yes or no. Instead, it reframes the question itself by examining what it truly means to "invent," by understanding intelligence as an evolutionary continuum rather than a singular event, by situating India within the deep cognitive lineage of human thought, and by explaining why this inquiry shakes the foundations of modern intellectual history.

## What "Invent" Truly Means

In modern discourse, invention is often understood as the creation of a physical object or a discrete technological artifact. Under this definition, Artificial Intelligence would clearly belong to the age of computers, algorithms, and data centers. However, this narrow framing obscures a more fundamental truth: technologies are built upon ideas, and ideas have histories far older than their mechanical expressions.

If invention is defined only as the moment a machine is assembled, then the intellectual scaffolding that made the machine possible disappears from view. Writing would be credited to the printing press rather than to language. Mathematics would begin with calculators rather than with numbers. By this logic, intelligence itself would have no past before computers, an assumption that collapses under scrutiny.

A broader and more accurate understanding of invention recognizes foundational contributions:

- The creation of conceptual systems
- The formalization of rules and logic
- The abstraction of processes from human action
- The transmission of structured reasoning across generations

When viewed through this lens, Artificial Intelligence is not an invention in the conventional sense but a culmination. It is the mechanization of ways of thinking that humans developed long before machines existed. The question then becomes not who built the first AI system, but who first conceived the principles that make intelligent systems possible.

## Intelligence as Evolution, Not Event

Artificial Intelligence did not emerge suddenly in 1956 at the Dartmouth Conference, nor did it originate solely with digital computers. That moment marked a naming and formalization, not a birth. Intelligence artificial or otherwise has evolved through centuries of abstraction, systematization, and symbolic representation.

Human civilizations have always sought to externalize intelligence: through language, mathematics, writing, law, ritual, and logic. Each step moved cognition from the individual mind into shared systems. Algorithms existed long before the word itself; they lived in procedures, sutras, and rules for calculation, grammar, astronomy, and decision-making.

Seen this way, AI represents the latest phase in a long evolutionary process:

- From oral reasoning to written formalism
- From intuition to rule-based systems
- From human execution to mechanical execution
- From symbolic manipulation to automated inference

India's ancient intellectual traditions were particularly advanced in this evolutionary arc. They did not merely record observations; they formalized reasoning itself. They treated knowledge as something that could be compressed, transmitted, executed, and verified core attributes of algorithmic thought.

To frame intelligence as an event rather than an evolution is to misunderstand its nature. Artificial Intelligence is not a rupture from the past; it is a continuation of humanity's oldest project: making thought systematic.

## India's Role in the Cognitive Lineage

India's contribution to this lineage lies not in machines, but in mental architectures. Long before modern computing, Indian scholars developed systems that displayed key characteristics of artificial intelligence:

abstraction, recursion, prediction, optimization, and rule-governed transformation.

The concept of Śūnya enabled positional notation, without which computation at scale is impossible. Panini's grammar demonstrated how a finite set of rules could generate infinite expressions a principle mirrored in modern programming languages and AI language models. Aryabhata and Bhāskarāchārya built predictive mathematical models that separated calculation from the calculator. Brahmagupta formalized operations on zero and negatives, laying groundwork for algebraic reasoning.

These were not isolated achievements. They formed an integrated epistemic culture in which intelligence was treated as something that could be structured independently of individual minds. This is the defining characteristic of artificial intelligence.

Crucially, these ideas did not remain confined to India. Through translation movements and knowledge exchanges, they flowed into the Islamic world and then into Europe, where they were absorbed into the foundations of modern science and computation. The lineage is continuous, even if its origins were obscured.

India's role, therefore, is not that of a forgotten inventor of machines, but of a foundational architect of the intellectual conditions that made intelligent machines conceivable.

## Why the Question Shakes the World

Asking whether India invented Artificial Intelligence unsettles more than historical timelines. It challenges deeply embedded assumptions about where knowledge comes from and who is entitled to define progress. Modern scientific narratives have often positioned Europe and later the West as the primary engines of rationality and innovation, while other civilizations are cast as spiritual, mystical, or pre-scientific.

This question disrupts that hierarchy. It suggests that some of the most "modern" ideas formal systems, abstraction, algorithmic thinking have ancient, non-Western roots. For institutions built on linear narratives of progress, this is profoundly destabilizing.

The discomfort also arises because Artificial Intelligence is not merely a technology; it is power. Whoever defines its origins shapes its future. Acknowledging India's role forces a reconsideration of intellectual sovereignty, ethical frameworks, and global leadership in AI governance.

It also exposes a deeper anxiety: if intelligence itself has a shared civilizational history, then no single culture can claim moral ownership of its outcomes. This complicates narratives of technological destiny and control.

The question shakes the world because it reveals that Artificial Intelligence is not just a technical achievement, but a mirror reflecting humanity's collective past and its unresolved biases.

# *Chapter 26*

# From Śūnya to the Singularity

The journey from Śūnya to the singularity is not merely a technological arc; it is a civilizational story of how humanity has understood intelligence itself. Śūnya, the concept of zero, was never simply a numeral. It was a philosophical breakthrough a recognition that absence could be represented, that nothingness could be structured, and that emptiness could hold meaning. From this conceptual leap emerged the foundations of modern computation. The singularity, by contrast, represents a speculative horizon: a future point where artificial intelligence surpasses human cognitive capabilities. Between these two poles lies a continuum of ideas, values, and choices that will define the next era of intelligence.

Understanding this continuum requires abandoning the myth that intelligence suddenly emerged with machines. Intelligence has always been embedded in systems mathematical, linguistic, ethical, and social. What has changed is the scale and speed at which intelligence now operates. As we stand at the threshold of a new epoch, the lessons encoded in ancient knowledge systems become increasingly relevant.

## From Ancient Zero to Future Intelligence

Śūnya was revolutionary because it allowed systems to breathe. Without zero, place-value notation collapses, arithmetic loses coherence, and computation becomes rigid. Zero enabled flexibility, abstraction, and scalability qualities that define modern algorithms. Every binary system, every digital circuit, and every AI model is built upon the presence of zero as both a placeholder and a boundary.

Yet Śūnya was never treated as a mere technical tool. In Indian philosophy, it symbolized potential, balance, and continuity. It represented the space from which form arises and into which form dissolves. This dual understanding technical and metaphysical gave Indian mathematics a depth that extended beyond calculation.

Future intelligence, whether artificial or hybrid, will require a similar balance. As AI systems grow more complex, they demand frameworks that account not only for performance but for meaning. Intelligence divorced from context risks becoming brittle, misaligned, or destructive.

The path from zero to advanced intelligence is not linear. It is recursive. Each leap in abstraction requires a grounding principle that maintains coherence. Śūnya provided that grounding for mathematics. Ethical and philosophical anchors must provide it for future intelligence.

## Human–Machine Co-Evolution

The dominant narrative around artificial intelligence often frames machines as replacements for humans. This narrative is both misleading and dangerous. Intelligence has never evolved in isolation. Human cognition itself is shaped by tools language, writing, symbols, and technologies that extend memory and reasoning.

Artificial intelligence represents the next phase of this co-evolution. Machines amplify human capacities, but they also reshape how humans think, decide, and interact. Navigation systems alter spatial reasoning. Recommendation engines influence taste and preference. Predictive analytics shift decision-making from intuition to probability.

This co-evolution raises profound questions about agency and identity. If machines increasingly anticipate our needs and behaviors, where does human choice end and algorithmic suggestion begin? The answer lies not in rejecting technology, but in designing systems that preserve reflective space.

Ancient traditions emphasized self-awareness as the foundation of intelligence. Tools were meant to serve human purpose, not replace human judgment. In this framework, co-evolution is guided by intentionality.

Human–machine co-evolution must therefore be symbiotic rather than competitive. Machines excel at scale, speed, and pattern recognition. Humans excel at meaning-making, moral reasoning, and contextual judgment. The future of intelligence depends on preserving this complementary relationship.

## The Next 100 Years of Intelligence

The next century will redefine intelligence in ways that are difficult to predict, but certain trajectories are already visible. AI systems will become more integrated into everyday life, operating invisibly across healthcare, governance, education, and infrastructure. Intelligence will shift from isolated tools to ambient environments.

Several developments are likely to shape this future:

- Intelligence will become decentralized, embedded in networks rather than centralized systems
- Human–AI collaboration will replace pure automation in critical domains
- Ethical governance will become a competitive advantage, not a constraint
- Cultural frameworks will influence how intelligence is deployed and regulated

The most significant challenge will not be technological, but moral. As systems gain predictive power, societies must decide how much autonomy to delegate. The temptation to optimize efficiency at the cost of freedom will grow.

The concept of the singularity suggests a moment of irreversible acceleration. Yet history teaches that intelligence evolves through adaptation, not rupture. Every transformative technology has required periods of recalibration. The printing press reshaped thought, but it also demanded new norms around authorship and truth. AI will demand similar recalibration.

The future of intelligence will be determined less by hardware breakthroughs and more by philosophical clarity. Societies that integrate wisdom into design will navigate this transition more successfully than those driven solely by competition.

## Why Bharat's Past Shapes the World's Future

Bharat's contribution to the history of intelligence lies not only in inventions, but in perspectives. The integration of mathematics with philosophy, logic with ethics, and knowledge with responsibility offers a model uniquely suited to the challenges of artificial intelligence.

Indian knowledge systems treated intelligence as relational rather than extractive. Knowledge was meant to harmonize with nature and society. This contrasts with modern paradigms that prioritize domination and control.

As the world grapples with AI ethics, governance, and long-term impact, these perspectives gain renewed relevance. Concepts such as Dharma emphasize duty over profit. Karma highlights consequence over intent. Ahimsa prioritizes harm reduction over efficiency. These ideas offer a vocabulary for ethical intelligence that transcends cultural boundaries.

Bharat's past also demonstrates the value of openness. Knowledge flowed across borders through translation and dialogue. This tradition counters contemporary fears of technological isolationism and suggests a path toward collaborative intelligence.

The future will belong to civilizations that combine technical excellence with moral depth. Bharat's historical experience positions it as a custodian of this synthesis.

# Epilogue

## The New Civilization of Intelligence

Human history has always been shaped by the tools we create, but only rarely do those tools reshape how intelligence itself is expressed. Artificial intelligence marks one of those rare inflection points. It is not merely a technological advancement; it is the emergence of a new civilizational layer where decision-making, prediction, creativity, and coordination increasingly occur through machines. This shift does not replace humanity's past it amplifies it. The question facing the world is not whether this new civilization of intelligence will emerge, but what values will guide it.

For the first time, intelligence is no longer confined to biological minds. It is distributed across networks, encoded in algorithms, and embedded in systems that operate at scales far beyond individual perception. Yet intelligence without wisdom is fragile. History offers countless examples of societies that mastered power but failed to master responsibility. The coming era will test whether humanity has learned from those lessons.

This new civilization of intelligence demands a redefinition of progress. Speed, efficiency, and optimization are no longer sufficient metrics. The true measure of advancement lies in whether intelligence deepens understanding, reduces suffering, and enhances human dignity. Technology that accelerates inequality, erodes agency, or obscures accountability may be impressive, but it cannot be considered civilizationally mature.

India's intellectual traditions offer a reminder that intelligence has always been more than calculation. From śūnya to structured reasoning, from grammar to astronomy, intelligence was understood as a disciplined harmony between logic and meaning. The future of artificial intelligence will depend on rediscovering that balance. Machines can process information, but only humans can assign purpose.

## A Message to Future Builders

To those who will design, train, deploy, and govern intelligent systems, the responsibility is unprecedented. You are not merely building tools; you are shaping the cognitive infrastructure of civilization. Every model encodes assumptions. Every dataset reflects values. Every system influences behavior. Neutrality is an illusion.

Future builders must resist the temptation to equate capability with legitimacy. Just because something can be automated does not mean it should be. Intelligence must remain accountable to human judgment, ethical reflection, and social consequence. Transparency, humility, and restraint are not obstacles to innovation they are its safeguards.

The next century will be defined by how well humanity integrates machine intelligence without surrendering moral agency. The most successful systems will not be those that dominate humans, but those that elevate them. The most enduring achievements will not be measured in computational power, but in trust, fairness, and wisdom.

A new civilization of intelligence is taking shape. Its foundations are being laid now, line by line, decision by decision. The future does not belong to machines alone. It belongs to those who choose to build intelligence with conscience, power with purpose, and progress with responsibility.

# Appendices

The appendices serve as a bridge between the conceptual arguments of this book and the historical, mathematical, and linguistic evidence that underpins them. They are intended to provide readers with concrete reference points, organized data, and explanatory tools that make the narrative accessible and verifiable. While the main chapters focus on narrative, analysis, and philosophical interpretation, these appendices anchor the discussion in evidence, chronology, and terminological clarity. Together, they highlight the continuity of Indian intellectual traditions and their contributions to the evolution of computation and artificial intelligence.

## Chronological Map of Indian Scholars

Indian mathematics, astronomy, linguistics, and logic evolved over millennia, producing a lineage of scholars whose insights continue to inform modern computation and artificial intelligence. Mapping these figures chronologically allows us to trace the development of ideas from abstraction to application.

Aryabhata (476–550 CE) stands as a foundational figure. His work in mathematics and astronomy introduced algorithmic thinking to Indian intellectual life. Aryabhata's treatment of zero as both placeholder and concept enabled systematic calculations in arithmetic, algebra, and trigonometry. His predictive models for planetary movements demonstrated a computational mindset centuries before mechanical computing existed.

Brahmagupta (598–668 CE), building on Aryabhata's foundation, formalized operations with zero and negative numbers. His work in algebra, particularly in "Brahmasphutasiddhanta," included rules that closely resemble early formal systems, allowing calculation with previously undefined quantities.

Panini (circa 500 BCE), often cited for his grammatical innovations, developed the Ashtadhyayi, a rule-based formal system for Sanskrit. While primarily linguistic, Panini's use of meta-rules, recursion, and generative grammar mirrors modern programming language design and compiler theory. The precision and completeness of his framework suggest that algorithmic thinking was embedded in Indian intellectual culture long before the invention of electronic computers.

Bhāskarāchārya (1114–1185 CE) extended mathematical reasoning into predictive modeling, optimization, and early calculus concepts. His texts, Lilavati and Bijaganita, illustrate systematic approaches to problem-solving that combine arithmetic, algebra, and geometry. The ability to model natural phenomena mathematically reflects a mindset aligned with the later goals of artificial intelligence: abstraction, prediction, and optimization.

Other notable figures, including Varahamihira (505–587 CE), Madhava (circa 1340–1425 CE), and Hemachandra (1089–1172 CE), contributed to trigonometry, series expansions, and combinatorial logic. Each scholar represents a node in the continuum of knowledge, demonstrating both specialization and the integration of mathematics with astronomy, linguistics, and philosophy.

This chronological map underscores that Indian intellectual achievements were not isolated events but a continuous thread of structured reasoning, abstraction, and practical application a tradition that provided the conceptual scaffolding for modern computation and AI.

## Knowledge Transmission Evidence Tables

One of the central arguments in this book is that Indian knowledge influenced global mathematics and computation. Evidence for this transmission is both textual and historical. Knowledge transmission tables compile these connections across centuries.

| Indian Scholar | Concept / Invention | Transmission Route | Influenced Figure / Region |
|---|---|---|---|
| Aryabhata | Place-value system, zero | Sanskrit → Arabic | Al-Khwarizmi, Persian mathematicians |
| Brahmagupta | Rules for zero and negatives | Sanskrit → Arabic → Latin | European algebraic development, Fibonacci |

| Indian Scholar | Concept / Invention | Transmission Route | Influenced Figure / Region |
|---|---|---|---|
| Panini | Formal grammar, recursive rules | Sanskrit → Arabic → Europe (via translation) | Modern compiler theory, formal linguistics |
| Bhāskarāchārya | Predictive astronomy, combinatorics | Sanskrit manuscripts, commentary schools | European astronomy and calculus precursors |
| Madhava | Infinite series, trigonometry | Kerala manuscripts → Jesuit scholars | Leibniz, Gregory, European series expansion methods |

These tables highlight both direct and indirect pathways. Indian texts were translated into Arabic during the medieval period, then into Latin and other European languages. The formal structures embedded in these texts informed algebraic, arithmetic, and computational traditions that ultimately shaped modern science. By organizing this information systematically, readers can appreciate the evidence of continuity rather than isolated coincidence.

## Mathematical Proof Lineage

Mathematical rigor in India predates the formalization of proof structures in Europe. Indian mathematicians often embedded proofs implicitly through examples, logical reasoning, and algorithmic processes. A lineage of proofs demonstrates the evolution of abstraction and formal reasoning:

- Aryabhata: Proofs were primarily algorithmic and empirical, using stepwise calculation for astronomical prediction. While not "proof" in the modern formal sense, these methods relied on reproducibility and predictive accuracy.

- Brahmagupta: Introduced generalized rules for operations with zero and negative numbers. His logical deductions resemble axiomatic structures, where outcomes are derived systematically from defined rules.

- Bhāskarāchārya: Combined deductive and inductive reasoning. For instance, solutions to quadratic and cubic equations were validated through iterative methods, illustrating early algorithmic verification techniques.

- Madhava: Series expansions for sine, cosine, and arctangent functions were accompanied by error estimation methods, foreshadowing rigorous convergence proofs in later European calculus.

The lineage reveals continuity: from empirical verification to abstraction and error estimation. Indian mathematical reasoning was not isolated arithmetic; it was a structured intellectual ecosystem designed to model reality, predict outcomes, and generalize rules a direct precursor to the logic underlying AI algorithms.

## AI Glossary for General Readers

To bridge historical concepts with modern AI, this appendix provides a glossary of terms, ensuring accessibility for general readers while connecting ancient Indian knowledge with contemporary AI vocabulary.

- Algorithm: Stepwise procedure for calculation or problem-solving. Ancient equivalent: Aryabhata's astronomical computations and Panini's grammatical rules.
- Binary Logic: Representation of two states (0 and 1). Ancient equivalent: Conceptualization of Śūnya and existence.
- Predictive Modeling: Using historical data to forecast future outcomes. Ancient equivalent: Bhāskarāchārya's planetary simulations.
- Neural Network: Computational model inspired by brain structures. Ancient equivalent: Systemic thinking in Indian philosophy, integrating pattern recognition, memory, and inference.
- Optimization: Process of finding the best solution under constraints. Ancient equivalent: Bhāskarāchārya's calculations for minimal or maximal outcomes in astronomy or combinatorics.
- Recursion: Rule or function that references itself. Ancient equivalent: Panini's grammar rules applying operations iteratively.

- Formal System: A set of rules and symbols to generate valid statements. Ancient equivalent: Ashtadhyayi's grammatical structures.

This glossary emphasizes continuity between ancient and modern thought, demonstrating that AI's conceptual scaffolding is embedded in historical intellectual traditions.

## Sanskrit–Math–AI Term Mapping

Understanding AI through the lens of Indian knowledge systems requires recognizing terminological correspondences. Sanskrit terminology often encodes computational and logical principles:

- Śūnya (शून्य) → Zero, placeholder, null value, foundational abstraction in computation
- Anu (अणु) → Smallest unit, analogous to bits or discrete data points
- Sutra (सूत्र) → Rule or formula, akin to algorithmic instruction
- Vritti (वृत्ति) → Pattern or transformation, used in predictive modeling
- Pramana (प्रमाण) → Measure, validation, or proof, akin to empirical verification in AI systems
- Vyavaharika (व्यवहारिक) → Practical application, analogous to applied AI solutions
- Bhavana (भवना) → Conceptual imagination or modeling, used in theoretical abstraction

Mapping these terms reveals a conceptual continuity: Indian scholars developed a sophisticated symbolic and operational vocabulary century before modern computation. Their frameworks were both philosophical and technical, combining abstraction with systematic application. Understanding these correspondences helps modern AI practitioners and historians appreciate the depth of historical contributions.

# Bibliography & Academic References

The study of intelligence both human and artificial requires an interdisciplinary lens that spans centuries, civilizations, and modes of knowledge. This bibliography and reference section consolidates the foundational sources, manuscripts, academic research, and policy frameworks that underpin the arguments and historical reconstructions presented in this book. By providing structured references across Indian primary sources, Arabic manuscripts, European scientific archives, modern AI research, and government policy reports, this section emphasizes both the depth and breadth of the scholarship that informs the narrative. It also highlights the lineage of knowledge transmission from India to the wider world, bridging historical scholarship with contemporary AI discourse.

## Indian Primary Sources

Indian scholarship has historically been characterized by meticulous documentation, logical rigor, and a systematization of knowledge across mathematics, astronomy, linguistics, and philosophy. The primary sources referenced in this work form the core of the civilizational argument for India's foundational contributions to computation and intelligence. They include original Sanskrit manuscripts, commentaries, and treatises that illustrate both theoretical innovation and practical application.

- Aryabhatiya by Aryabhata (476 CE): Aryabhata's seminal text on mathematics and astronomy establishes the first systematic algorithmic approach to celestial calculations. It contains trigonometric methods, place-value notation, and predictive models of planetary motion, laying the groundwork for algorithmic reasoning.
- Brahmasphutasiddhanta by Brahmagupta (598–668 CE): This treatise formalizes arithmetic operations on zero, negative numbers,

and rational quantities. It serves as an early example of formalized computation and system-based error checking.

- Ashtadhyayi by Panini (circa 500 BCE): Panini's grammar codifies Sanskrit into a rule-based system that incorporates recursion, meta-rules, and generative structures. This work has influenced modern computational linguistics and formal programming languages.
- Lilavati and Bijaganita by Bhāskarāchārya (1114–1185 CE): These texts demonstrate predictive modeling, optimization, and combinatorial reasoning applied to both mathematical and astronomical problems.
- Varahamihira's Pancha-Siddhantika (6th century CE): A compendium of astronomical knowledge integrating multiple computational approaches.
- Other Key Manuscripts: Works by Madhava of Kerala (infinite series and early calculus), Hemachandra (combinatorics and numerical patterns), and Ganita-Sarasangraha (compendiums of mathematical algorithms) provide additional insights into systematic reasoning and predictive computation in classical India.

These primary sources reflect a continuous tradition of structured reasoning, abstraction, and applied intelligence, offering clear evidence of India's early engagement with concepts foundational to computation and artificial intelligence.

## Arabic Manuscripts

During the medieval period, Indian knowledge systems were transmitted to the Islamic world, primarily through translation efforts in centers such as Baghdad. These Arabic manuscripts played a pivotal role in disseminating Indian arithmetic, algebra, and astronomical reasoning to Europe and beyond.

- Al-Khwarizmi's Kitab al-Jabr wa-l-Muqabala (circa 820 CE): While often cited as the origin of algebra, Al-Khwarizmi's work is heavily influenced by Indian numerals, arithmetic methods, and algorithmic procedures. Indian sources on zero and decimal systems informed his development of computational rules.

- Arabic Astronomical Treatises: Texts such as Al-Biruni's India and other commentaries translated Sanskrit astronomical works, preserving and interpreting Aryabhata, Brahmagupta, and Bhāskarāchārya for wider audiences.
- Transmission Evidence: Manuscripts housed in libraries in Istanbul, Cairo, and Tehran show direct cross-references to Indian computation methods, illustrating a lineage of knowledge transfer that shaped Islamic mathematics and, subsequently, European Renaissance mathematics.

These Arabic manuscripts serve as an intermediate node, connecting Indian conceptual innovations to global scientific development. They also provide insight into how Indian ideas were recontextualized in different linguistic, philosophical, and technological settings.

## European Scientific Archives

Europe's engagement with Indian mathematical knowledge primarily occurred during the Renaissance and early modern periods. European archives contain translations, commentaries, and treatises that demonstrate how Indian methods influenced Western arithmetic, algebra, and the development of early computing concepts.

- Fibonacci's Liber Abaci (1202 CE): Fibonacci's introduction of the Hindu-Arabic numeral system to Europe shows clear influence from Indian arithmetic transmitted via Arabic intermediaries.
- Leibniz's Binary Studies (17th century): Leibniz's formulation of binary arithmetic, which underpins modern computation, has conceptual parallels with the logic of Śūnya and duality in Indian mathematics.
- University and Library Archives: Collections in Oxford, Cambridge, Paris, and Florence preserve translations of Indian texts into Latin, providing evidence of intellectual transmission and adaptation. These

archives also contain marginalia, commentaries, and cross-references demonstrating European engagement with Indian formal systems.

- Scientific Correspondences and Treatises: Letters and manuscripts in European archives document debates on numerical systems, zero, and algorithmic methods, confirming India's indirect but profound influence on Western computation and mathematical thought.

These European sources provide external validation of the continuity of ideas from Indian civilization to the global scientific community, illustrating both adoption and reinterpretation.

## Modern AI Research Papers

Contemporary artificial intelligence research builds upon mathematical, linguistic, and algorithmic principles rooted in centuries-old scholarship. To contextualize the historical argument within present-day AI, this book references a curated set of academic papers from leading conferences, journals, and institutions.

- Neural Networks and Deep Learning: Papers documenting the development of perceptrons, backpropagation, convolutional neural networks, and transformer architectures. These works illustrate modern algorithmic design and predictive modeling, which conceptually echo Indian algorithmic reasoning.

- Natural Language Processing (NLP): Research on grammar-based models, language tokenization, and syntax-driven architectures demonstrates continuity with Panini's Ashtadhyayi, highlighting parallels between ancient linguistic formalism and AI language modeling.

- Predictive Analytics and Optimization: Studies on forecasting, time series analysis, and optimization algorithms reflect principles embedded in Bhāskarāchārya's mathematical modeling and early calculus.

- Ethics and AI Policy Research: Papers addressing bias, fairness, and explainability provide a framework for evaluating AI through ethical lenses reminiscent of Indian philosophical principles like Dharma and Ahimsa.

In referencing these modern research papers, the book bridges historical scholarship with cutting-edge technological practice, demonstrating that ancient conceptual innovations continue to resonate in contemporary AI frameworks.

## Government AI Policy Reports

Understanding India's modern re-entry into the AI landscape requires an examination of official policy documents. These reports offer insight into regulatory frameworks, strategic initiatives, and national priorities that guide AI development.

- National Strategy for Artificial Intelligence (NITI Aayog, 2018): Outlines India's vision for AI, emphasizing inclusive growth, ethical development, and sector-specific adoption.
- Digital India and AI Integration Reports: Documents that detail AI deployment across governance, finance, healthcare, and public service infrastructure.
- Global Policy Comparisons: Reports from the European Union (EU AI Act), United Nations (AI Ethics Principles), and OECD provide context for India's positioning in international AI governance.
- Defense, Space, and Health AI Policies: Policy papers demonstrating India's efforts to leverage AI for strategic autonomy and public welfare, reflecting a continuation of knowledge sovereignty that traces back to civilizational practices.

These policy reports illustrate that India is not merely participating in AI development but actively shaping its ethical, societal, and strategic contours.

# Index

The index of From Śūnya to Algorithms: Did India Invent Artificial Intelligence is designed not only to facilitate navigation of the text but also to highlight the interconnectedness of scholars, mathematical concepts, AI terminology, and civilizational contributions. By contextualizing each entry, this index functions as both a practical reference and a thematic guide, allowing readers to trace the intellectual lineage of ideas from ancient India to modern artificial intelligence.

## Scholars

The individuals highlighted in this work represent pivotal nodes in the historical and conceptual evolution of intelligence, computation, and predictive modeling. They are indexed not just by name, but also by their contributions, methodologies, and intellectual impact.

- Aryabhata (476–550 CE): Referenced across discussions of algorithmic astronomy, trigonometric computation, and the place-value system. Aryabhata's mathematical treatises reveal early algorithmic thinking, including techniques for predicting planetary positions and calculating eclipses. He is linked to the conceptual development of zero and structured calculation.
- Brahmagupta (598–668 CE): Indexed in the context of zero, negative numbers, and formal computational rules. His Brahmasphutasiddhanta codifies operations on abstract quantities, demonstrating a proto-algorithmic approach to mathematics. He is also associated with error correction and early system theory.
- Panini (circa 500 BCE): Recognized for the Ashtadhyayi, his grammar codex is a cornerstone for formal linguistic systems. Entries detail his use of recursion, meta-rules, and generative syntax,

all of which anticipate computational language modeling. Panini's contributions link to AI discussions, particularly natural language processing and language model architecture.

- Bhāskarāchārya (1114–1185 CE): Indexed for predictive modeling, optimization, and early calculus concepts. His works Lilavati and Bijaganita illustrate structured problem-solving, optimization, and combinatorial reasoning. References include his influence on time series forecasting and planetary simulations, bridging ancient reasoning with AI predictive analytics.
- Madhava of Kerala (circa 1340–1425 CE): Included for infinite series and early calculus contributions, particularly in trigonometric expansions. Madhava's work is linked to modern numerical approximation methods and algorithmic reasoning in AI.
- Varahamihira (505–587 CE): Indexed for his astronomical compendiums and practical computation strategies, including integration of multiple methodologies for prediction and observation.
- Al-Khwarizmi (circa 780–850 CE): Cited in discussions of algorithmic terminology and algebra. Though writing in the Arab world, his work demonstrates the influence of Indian computational systems and zero-based arithmetic.

By including scholars in the index with contextual references, readers can trace the continuity of ideas and observe the connections between individual contributions, civilizational knowledge, and modern computational paradigms.

## Mathematical Concepts

Mathematics in India provided the foundation for algorithmic thought and predictive modeling. This section of the index emphasizes both technical definitions and their intellectual lineage.

- Zero (Śūnya): Indexed in contexts ranging from philosophical abstractions in the Upanishads to practical arithmetic operations in

Aryabhata and Brahmagupta. Its conceptual depth underpins binary logic and digital computation.

- Place-Value System: Referenced in discussions of algorithmic efficiency and numerical representation. Aryabhata's use of positional notation is highlighted as a precursor to computational systems.
- Negative Numbers: Brahmagupta's formalization of operations with negatives, including additive inverses, is cross-referenced with error correction methods in computation.
- Recursion: Panini's grammatical rules employ recursion, meta-rules, and generative structures. This concept is indexed in relation to both language processing and algorithmic reasoning.
- Combinatorics and Series Expansion: Bhāskarāchārya and Madhava are indexed for their work on permutations, combinations, and trigonometric series. These mathematical constructs inform modern AI algorithms, particularly in optimization and numerical analysis.
- Optimization and Limit Theory: Indexed under Bhāskarāchārya's predictive modeling, showing early approaches to maximizing or minimizing outcomes, which are fundamental in machine learning and AI model design.
- Probability and Forecasting: Connected to planetary simulations, predictive analytics, and modern time-series AI models. Historical roots are traced from Indian astronomical calculations to contemporary predictive frameworks.
- Linear Algebra and Graph Theory: Cross-referenced with modern AI discussions, emphasizing how foundational Indian mathematics contributed to the computational tools necessary for machine learning and neural network design.

This indexing approach situates mathematical concepts within a continuous intellectual tradition, linking ancient knowledge directly to modern computational methodologies.

## AI Terms

Modern artificial intelligence terminology is indexed alongside historical and conceptual antecedents, creating a dialogue between ancient reasoning and contemporary technology.

- Algorithm: Cross-referenced with Aryabhata's computation, Panini's rule systems, and Al-Khwarizmi's algebraic methods. The index highlights both historical origins and current applications in AI.
- Neural Networks: Linked to ancient models of parallel reasoning, Vedic pattern recognition, and early ideas of interconnected processing. Entries point to both theoretical origins and contemporary implementations.
- Machine Learning / Predictive Analytics: Indexed alongside Bhāskarāchārya's forecasting methods, demonstrating historical continuity in prediction-based reasoning.
- Natural Language Processing (NLP): Associated with Panini's Ashtadhyayi, tokenization, and grammar-based modeling. References emphasize both linguistic structure and AI language modeling.
- Optimization: Cross-referenced between early calculus applications, combinatorial reasoning, and modern algorithmic learning models.
- Consciousness and AGI: Entries relate philosophical discussions of human awareness, cognitive models, and machine intelligence limitations. This includes references to Vedic and Upanishadic concepts of mind and cognition.
- Data Sovereignty: Linked to India's modern AI policies, emphasizing algorithmic independence, ethical AI, and the preservation of national knowledge assets.
- Ethical AI: Indexed across discussions of Dharma, Karma, and Ahimsa in relation to modern AI governance frameworks.

By including AI terms in the index with historical antecedents, the book underscores the continuity of computational ideas from classical India to the present.

## Civilizational References

This section captures broader socio-cultural and civilizational contexts that inform the history and philosophy of computation and intelligence.

- Gurukul System: Indexed under education and computational ecosystems, illustrating how knowledge transmission was structured for systemic learning.
- Takshashila and Nalanda: Cross-referenced with scholarship on predictive modeling, structured reasoning, and early algorithmic thinking. These universities exemplify applied knowledge systems in ancient India.
- Knowledge Transmission Routes: Indexed across India-to-Arab-to-Europe pathways, highlighting the movement and reinterpretation of Indian mathematics, astronomy, and linguistic systems.
- Colonial Knowledge Suppression: References include textbook whitewashing, erasure of non-Western contributions, and the politics of scientific ownership, emphasizing how civilizational credit was historically contested.
- Sanskrit Language: Indexed in connection with Panini, AI-ready linguistic structures, and modern natural language processing.
- Śūnya Civilization: Contextualized as a conceptual and philosophical foundation for structured reasoning, algorithmic thought, and abstraction.
- Global AI Ethics: Cross-referenced with Indian ethical principles, modern AI governance, and international frameworks. This connects ancient civilizational values to contemporary technology debates.
- Digital Public Infrastructure: Aadhaar, UPI, and other initiatives are indexed for their role in positioning India as a global player in AI, linking civilizational continuity to modern technological agency.

www.ingramcontent.com/pod-product-compliance
Lightning Source LLC
LaVergne TN
LVHW100307210926
848480LV00003B/1042
* 9 7 9 8 9 0 3 9 2 1 3 4 8 *